COLS
WRIGHT STATE UNIVERSITY
UNIVERSITY LIBRARY
0 00 13 00752845 5

D1560479

GRACE WAHBA
University of Wisconsin at Madison

Spline Models for Observational Data

SOCIETY FOR INDUSTRIAL AND APPLIED MATHEMATICS
PHILADELPHIA, PENNSYLVANIA 1990

Library of Congress Cataloging-in-Publication Data

Wahba, Grace, 1934–
Spline models for observational data / Grace Wahba
p. cm. — (CBMS-NSF Regional Conference series in applied mathematics ; 59)
Based on a series of 10 lectures at Ohio State University at Columbus, Mar. 23–27, 1987.
Includes bibliographical references (p.).
ISBN 089871-244-0
1. Spline theory. 2. Mathematical statistics. I. Title
II. Series
QA224.W34 1990 89-28687
519.5—dc20

Dedicated to the memory of my father,
Harry Goldsmith.

Contents

Foreword

This monograph is based on a series of 10 lectures at Ohio State University at Columbus, March 23–27, 1987, sponsored by the Conference Board of the Mathematical Sciences and the National Science Foundation. The selection of topics is quite personal and, together with the talks of the other speakers, the lectures represent a story, as I saw it in March 1987, of many of the interesting things that statisticians can do with splines. I told the audience that the priority order for topic selection was, first, obscure work of my own and collaborators, second, other work by myself and students, with important work by other speakers deliberately omitted in the hope that they would mention it themselves. This monograph will more or less follow that outline, so that it is very much slanted toward work I had some hand in, although I will try to mention at least by reference important work by the other speakers and some of the attendees. The other speakers were (in alphabetical order), Dennis Cox, Randy Eubank, Ker-Chau Li, Douglas Nychka, David Scott, Bernard Silverman, Paul Speckman, and James Wendelberger. The work of Finbarr O'Sullivan, who was unable to attend, in extending the developing theory to the non-Gaussian and nonlinear case will also play a central role, as will the work of Florencio Utreras.

Now, a bit of background is given. The (univariate, natural) polynomial spline $s(x) = s_n^m(x)$ is a real-valued function on $[a, b]$ defined with the aid of n so-called knots $-\infty \le a < x_1 < x_2 < \ldots < x_n < b \le \infty$ with the following properties: (i) $s \in \pi^{m-1}$, $x \in [a, x_1]$, $x \in [x_n, b]$, (ii) $s \in \pi^{2m-1}$, $x \in [x_i, x_{i+1}]$, $i = 1, \ldots, n-1$, (iii) $s \in C^{2m-2}$, $x \in (-\infty, \infty)$, where π^k is the polynomials of degree k or less, and C^k is the class of functions with k continuous derivatives. In words, $s(\cdot)$ is a piecewise polynomial in each interval $[a, x_1]$, $[x_i, x_{i+1}]$ $i = 1, 2, \ldots, n-1$, $[x_n, b]$ with the pieces joined at the knots so that s has $2m-2$ continuous derivatives. It takes m coefficients to define s to the left of x_1, m coefficients to define s to the right of x_n, and $(n-1)$ $2m$ coefficients to define s in the $(n-1)$ interior intervals for a total of $2mn$ coefficients. The continuity conditions (iii) provide $(2m-1)n$ coefficients and (it can be shown that) the values of $s(x)$ at the n points $(x_1, \ldots, x_n)$ then provide the remaining n coefficients to define $s(x)$ uniquely. Schoenberg (1964a, 1964b) considered the following problem. Find f in the Sobolev space W_m of functions with $m-1$ continuous derivatives and mth derivative square integrable, to minimize

$$\int_a^b (f^{(m)}(x))^2 \, dx \tag{0.0.1}$$

subject to (iv) $f(x_i) = f_i$, $i = 1, 2, \ldots, n$. He showed that, provided $n \geq m$, this minimizer was the unique natural polynomial spline satisfying (i)–(iv). He called this object a spline, due to its resemblance (when $m = 2$) to the mechanical spline used by draftsmen. The mechanical spline is a thin reedlike strip that was used to draw curves needed in the fabrication of cross sections of ships' hulls. Ducks or weights were placed on the strip to force it to go through given points, and the free portion of the strip would assume a position in space that minimized the (two-dimensional) bending energy. With $m = 2$, the quantity (0.0.1) is the (one-dimensional) curvature. The terminology "natural" comes from the fact that if (0.0.1) is replaced by

$$\int_{x_1}^{x_n} (f^{(m)}(x))^2 \, dx$$

in the minimization problem then the solutions to the two problems will coincide in $[x_1, x_n]$, with the solution to the latter problem satisfying the so-called Neumann or "natural" boundary conditions $f^{(j)}(x_1) = f^{(j)}(x_n) = 0$, $j = m, m+1, \ldots, 2m-1$.

Statisticians are generally interested in smoothing rather than interpolating data. Consider the data model

$$y_i = f(x_i) + \epsilon_i, \quad i = 1, 2, \ldots, n \tag{0.0.2}$$

where $\epsilon = (\epsilon_1, \ldots, \epsilon_n)' \sim \mathcal{N}(0, \sigma^2 I)$ and f is only known to be "smooth." Consider, as an estimate of f, the solution to the following minimization problem. Find f in W_m to minimize

$$\frac{1}{n} \sum_{i=1}^{n} (y_i - f(x_i))^2 + \lambda \int_a^b (f^{(m)}(x))^2 \, dx, \tag{0.0.3}$$

for some $\lambda > 0$. This expression represents a tradeoff between fidelity to the data, as represented by the residual sum of squares, and "smoothness" of the solution, as represented by the square integral of the mth derivative. It was again shown by Schoenberg that the minimizer is a natural polynomial spline. It no longer interpolates the noisy data $y = (y_1, \ldots, y_n)$, but smoothes it, with the *smoothing parameter* λ controlling the tradeoff between fidelity and smoothness. It is data smoothing problems such as (0.0.3) and their myriad variations that we will be interested in, in this monograph.

Numerical analysts who were typically interested in exact data soon found wonderful things to do with spline functions, because of their ease of handling in the computer coupled with their good approximation theoretic properties. As an example of a good approximation theoretic property, suppose f is a function with $m-1$ continuous derivatives on $[x_1, x_n]$,

$\int_{x_1}^{x_n} (f^{(m)}(x))^2\,dx \le c^2$ and $\max|x_{i+1} - x_i| \le h$; then the natural spline of interpolation s_f that interpolates to f at $x_1, \ldots, x_n$, satisfies

$$\sup_{x \in [x_1, x_m]} |f^{(k)}(x) - s_f^{(k)}(x)| \le \text{const.}\ |h|^{m-k} \left[\int_a^b (f^{(m)}(x))^2\,dx \right]^{1/2},$$

$$k = 0, 1, \ldots, m-1,$$

with even higher rates of convergence if further conditions are imposed on f (see Schultz (1973a,b)). Other important references on splines from the point of view of numerical analysis with exact data are Golomb and Weinberger (1959), de-Boor and Lynch (1966), deBoor (1978), Schumaker (1981), Prenter (1975), and the conference proceedings edited by Greville (1968) and Schoenberg (1969). We will not concern ourselves with exact data here, however. Returning to noisy data, a discrete version of problems such as (0.0.3) were considered in the actuarial literature under the rubric graduation by Whittaker (1923), who considered smoothing $y_1, \ldots, y_n$ discretely by finding $f = (f_1, \ldots, f_n)$ to minimize

$$\frac{1}{n}\sum (y_i - f_i)^2 + \lambda \sum_{i=1}^{n-3} (f_{i+3} - 3f_{i+2} + 3f_{i+1} - f_i)^2.$$

Here we will primarily be concerned with estimating functions f defined on continuous index sets (as opposed to vectors) and will only mention in passing recent related work by Green (1985, 1987), Green and Yandell (1985), Green, Jennison, and Seheult (1983, 1985), and Steinberg (1983, 1984a,b) in the discrete setting. Steinberg's work relates some of the present results to the work of the English Bayesian school.

The first generalizations of (0.0.3) that are of interest concern the replacement of $\int_a^b (f^{(m)}(x))^2\,dx$ with more general quadratic penalty functionals, for example $\int_a^b (L_m f)^2(x)\,dx$, where L_m is an mth order differential operator satisfying some conditions, and the replacement of the *evaluation functionals* $f(x_i), i = 1, 2, \ldots, n$ by more general observational functionals $L_i f$, where L_i is a *bounded linear functional* on an appropriate space, for example,

$$L_i f = \int_a^b w_i(x) f(x)\,dx,$$

or

$$L_i f = f'(x_i).$$

Characterization of the solution to these generalized variational problems and more abstract variational problems was given in Kimeldorf and Wahba (1971). Historically that work is very close to that of Golomb and Weinberger (1959) and deBoor and Lynch (1966), and later work on characterizing solutions to variational problems arising in smoothing has, I believe, been made easier by the variational lemmas given there. George Kimeldorf and I also demonstrated the connection between these variational problems and Bayes estimates in Kimeldorf and Wahba (1970a,b), thus leading the way to showing that smoothing splines possess a double whammy—good approximation theoretic properties for models (0.0.2) and its generalizations, both when f is a fixed "smooth" function with certain properties, and when f is considered to be a sample function from a stochastic process. This connection between Bayes estimation and variational problems has its historical roots in the work of Parzen (1962, 1970). This work was done when George and I were both visitors at the Mathematics Research Center at the University of Wisconsin-Madison and we benefited from the stimulating atmosphere created there by Profs. Schoenberg, Greville, deBoor, and numerous visitors, including Larry Schumaker.

The formulas in Kimeldorf and Wahba (1971) were not very well suited to the computing capabilities of the day (they were essentially impractical) and the work did not attract much attention from statisticians. In fact offshoots of these papers were rejected in the early 1970s by mainstream statistics journals as being too "far out." Possibly the first spline paper in a mainstream statistics journal is the paper on histosplines by Boneva, Kendall, and Stefanov (1971). That paper lacks a certain rigor but certainly is of historical importance. Even well into the mid 1970s not very much was seen about splines in the statistics literature although the approximation theoretic literature was growing by leaps and bounds. In the later 1970s a number of things happened to propel splines to a popular niche in the statistics literature—computing power became available, which made the computation of splines with large data sets feasible, and, later, inexpensive; a good data-based method for choosing λ became available, and most importantly, splines engaged the interest of a number of creative researchers, notably including Bernard Silverman and some of the other speakers. Simultaneously the work of Duchon (1977), Meinguet (1979), Utreras (1979), Wahba and Wendelberger (1980), and others on multivariate thin-plate splines led to the development of a practical multivariate smoothing method, which (unlike the univariate spline) had few real competitors in the so-called "nonparametric curve smoothing" literature. Thin plate splines and "kriging" estimates, another multivariate smoothing method, are closely related (see Section 2.4 and Chapter 3). There rapidly followed splines and vector splines on the sphere, partial splines and interaction splines, variational problems where the data are non-Gaussian and where the observation functionals are nonlinear, and where linear inequality constraints are known to hold. Along with these generalizations came improved numerical methods, publicly available efficient software, numerous results in good and optimal theoretical properties, confidence statements and diagnostics, and many interesting and important applications. The body of spline methods available and under development provide a rich family of estimation and model building techniques that have found use in many scientific disciplines. Today it is hard to open an issue of the *Journal of the American Statistical Association*, the *Annals of Statistics*, or the *Journal of the Royal Statistical Society* without finding the word "spline" somewhere. It is certainly a pleasure to be associated with such a blossoming and important area of research.

The variational problems that we will be discussing can be treated from a unified point of view as optimization problems in a *reproducing kernel Hilbert space.* We will assume that the reader has a knowledge of the basic properties of Hilbert spaces of real-valued functions including the notion of *bounded linear functional* and the *Riesz representation theorem.* This background can be obtained by reading the first 39 pages of Akhiezer and Glazman (1963) or the first 34 pages of Halmos (1957). We will review the properties of reproducing kernels (r.k.'s) and reproducing kernel Hilbert spaces (r.k.h.s.'s) that we need. For further background the reader may refer to Aronszajn (1950). Although many of the results here can be obtained without the Hilbert space machinery (in particular, Schoenberg used mainly integration by parts), the fact that the solution to all of the quadratic optimization problems we discuss can be characterized in terms of a relevant reproducing kernel saves one from proving the same theorems over and over in different contexts. The r.k. in its reincarnation as a covariance also provides the crucial link to Bayesian estimation. Under a variety of circumstances the convergence rates of the various estimates can be related to the rate of decay of the eigenvalues of the reproducing kernel and the Fourier–Bessel coefficients of the function being estimated with respect to the eigenfunctions. I trust that the manuscript will be accessible to a second or third year graduate student in statistics who has read the aforementioned parts of Akhiezer and Glazman (1963) or Halmos (1957). I would like to assure the reader that the effort to master the basic properties of r.k.h.s., which regrettably are not part of the graduate statistics curriculum at many institutions, will be worth the effort.

All of the splines that we discuss in this book may be obtained as solutions to variational problems. We remark that there is a rich theory of multivariate piecewise polynomial splines that do not arise naturally as the solution to a variational problem. These splines are beyond the scope of the present work. The reader is referred to Hollig (1986) or Chui (1988).

The conference (and by extension, this monograph) was made possible by Prof. Sue Leurgans, who persuaded the National Science Foundation to sponsor the conference and whose superb organizational skills induced many of the active researchers in the field to attend; some came from great distances. She provided a flawless ambiance for scientific interaction, and made sure that the speaker got her homework done in time. Thanks are also due to the warm hospitality of Prof. and Mrs. Jagdish Rustagi, and other members of the Statistics Department at Ohio State University.

Of course very few researchers work in a vacuum, and the work of mine that is presented here owes much to nearly 20 years of a stimulating and supportive environment in the Statistics Department at the University of Wisconsin-Madison, and before that as a graduate student at Stanford University. Manny Parzen, my thesis advisor there, will recognize some of the introductory material from his class notes. For many years my research in splines was supported by the Office of Naval Research while Ed Wegman was Director of Probability and Statistics and then Director of Mathematics there. While managing a large program, Ed himself made some important contributions to the development of splines in statistics while prodding me on (see Wegman and Wright (1983)). My work on splines has more recently been supported by the National Science Foundation, and at present by the Air Force Office of Scientific Research. This support has been, and continues to be, invaluable.

September 1987

Unfortunately, we must tell the dear reader that many months passed from the writing of the first draft until the completion of the book. In that time the literature has continued to increase at an impressive rate. We have included brief mention of a few important topics for which results have become available since the CBMS conference. These include system identification, interaction splines and numerical methods for multiple smoothing parameters, and experimental design with noisy data (Sections 9.7, 10.1, 10.2, 11.3, 11.4, and 12.2).

I thank my patient typist, Thomas F. Orowan, S. Gildea and P. Cheng, who graciously provided LaTex macros, Jim Bucklew, Su-Yun Chen, Zehua Chen, Feng Gao, Chong Gu, Bill Studden, and Dave Reboussin, who helped proofread, and C. David Callan, for listening.

April 1989

CHAPTER 1

Background

1.1 Positive-definite functions, covariances, and reproducing kernels.

We begin with a general index set $\mathcal{T}$. Examples of $\mathcal{T}$ that are of interest follow:

$$\begin{aligned}
\mathcal{T} &= (1, 2, \ldots, N) \\
\mathcal{T} &= (\ldots, -1, 0, 1, \ldots) \\
\mathcal{T} &= [0, 1] \\
\mathcal{T} &= E^d \qquad \text{(Euclidean } d\text{-space)} \\
\mathcal{T} &= \mathcal{S} \qquad \text{(the unit sphere)} \\
\mathcal{T} &= \text{the atmosphere (the volume between two concentric spheres),}
\end{aligned}$$

etc. The text below is generally written as though the index set were continuous, but the discrete examples are usually special cases. A symmetric, real-valued function $R(s, t)$ of two variables $s, t \in \mathcal{T}$ is said to be *positive definite* if, for any real $a_1, \ldots, a_n$, and $t_1, \ldots, t_n \in \mathcal{T}$

$$\sum_{i,j=1}^{n} a_i a_j \, R(t_i, t_j) \geq 0,$$

and *strictly positive definite* if "$>$" holds. If $R(\cdot, \cdot)$ is positive definite, then we can always define a family $X(t)$, $t \in \mathcal{T}$ of zero-mean Gaussian random variables with covariance function R, that is,

$$E \, X(s)X(t) = R(s, t), \qquad s, t \in \mathcal{T}. \tag{1.1.1}$$

All functions and random variables in this book will be real valued unless specifically noted otherwise.

The existence of such a well-defined family of random variables in the continuous case is guaranteed by the Kolmogorov consistency theorem (see, e.g., Cramer and Leadbetter (1967, Chap. 3)). Given a positive-definite function $R(\cdot, \cdot)$ we are going to associate with it a *reproducing kernel Hilbert space* (r.k.h.s.). A (real) r.k.h.s. is a Hilbert space of real-valued functions on $\mathcal{T}$ with the property that, for each $t \in \mathcal{T}$, the evaluation functional L_t, which associates

f with $f(t)$, $L_t f \to f(t)$, is a bounded linear functional. The boundedness means that there exists an $M = M_t$ such that

$$|L_t f| = |f(t)| \le M\|f\| \quad \text{for all } f \text{ in the r.k.h.s.},$$

where $\|\cdot\|$ is the norm in the Hilbert space. We remark that the familiar Hilbert space $\mathcal{L}_2[0,1]$ of square integrable functions on $[0,1]$ does not have this property, no such M exists, and, in fact, elements in $\mathcal{L}_2[0,1]$ are not even defined pointwise.

If $\mathcal{H}$ is an r.k.h.s., then for each $t \in \mathcal{T}$ there exists, by the Riesz representation theorem, an element R_t in $\mathcal{H}$ with the property

$$L_t f = < R_t, f > = f(t), \qquad f \in \mathcal{H}. \tag{1.1.2}$$

R_t is called the representer of evaluation at t. Here, and elsewhere, we will use $< \cdot, \cdot >$ for the inner product in a reproducing kernel space. This inner product will, of course, depend on what space we are talking about. This leads us to the following theorem.

THEOREM 1.1.1. *To every* r.k.h.s. *there corresponds a unique positive-definite function (called the reproducing kernel* (r.k.)) *and conversely, given a positive-definite function R on $\mathcal{T} \times \mathcal{T}$ we can construct a unique* r.k.h.s. *of real-valued functions on $\mathcal{T}$ with R as its* r.k.

The proof is simple. If $\mathcal{H}$ is an r.k.h.s., then the r.k. is $R(s,t) = < R_s, R_t >$, where for each s, t, R_s and R_t are the representers of evaluation at s and t. $R(\cdot,\cdot)$ is positive definite since, for any $t_1, \ldots, t_n \in \mathcal{T}$, $a_1, \ldots, a_n$,

$$\begin{aligned} \sum_{i,j} a_i a_j \; R(t_i, t_j) &= \sum_{i,j} a_i a_j < R_{t_i}, R_{t_j} > \\ &= \|\Sigma a_j R_{t_j}\|^2 \ge 0. \end{aligned}$$

Conversely, given R we construct $\mathcal{H} = \mathcal{H}_R$ as follows. For each fixed $t \in \mathcal{T}$, denote by R_t the real-valued function with

$$R_t(\cdot) = R(t, \cdot). \tag{1.1.3}$$

By this is meant: R_t is the function whose value at s is $R(t,s)$. Then construct a linear manifold by taking all finite linear combinations of the form

$$\sum_i a_i R_{t_i} \tag{1.1.4}$$

for all choices of n, $a_1, \ldots, a_n$, $t_1, \ldots, t_n$ with the inner product

$$< \Sigma a_i \; R_{t_i}, \; \Sigma b_j \; R_{s_j} > = \sum_{ij} a_i b_j < R_{t_i}, R_{s_j} > = \Sigma a_i b_j R(t_i, s_j).$$

This is a well-defined inner product, since R is positive definite, and it is easy to check that for any f of the form (1.1.4) $< R_t, f > = f(t)$. In this linear manifold, norm convergence implies pointwise convergence, since

$$|f_n(t) - f(t)| = | < f_n - f, R_t > | \le \|f_n - f\| \; \|R_t\|.$$

Thus, to this linear manifold we can adjoin all limits of Cauchy sequences of functions in the linear manifold, which will be well defined as pointwise limits. The resulting Hilbert space is seen to be the r.k.h.s. $\mathcal{H}_R$ with r.k. R.

R is called the reproducing kernel, since

$$< R_s, R_t >=< R(s,\cdot),\ R(t,\cdot) >= R(s,t).$$

We will frequently denote by $\mathcal{H}_R$ the r.k.h.s. with r.k. R, and its inner product by $< \cdot,\cdot >_R$ or just $< \cdot,\cdot >$ if it is clear which inner product is meant. As a positive-definite function, under some general circumstances, R has an eigenvector-eigenvalue decomposition that generalizes the eigenvector-eigenvalue decomposition of a positive-definite matrix Σ, $\Sigma = \Gamma D \Gamma'$ with Γ orthogonal and D diagonal. Below we will show why the squared norm $\|f\|^2 = \|f\|_R^2$ can be thought of as a generalization of the expression $f'\Sigma^{-1}f$ with f a vector that appears in the multivariate normal density function with covariance Σ. In particular, suppose $R(s,t)$ continuous and

$$\int_{\mathcal{T}}\int_{\mathcal{T}} R^2(s,t)\,ds\,dt < \infty. \tag{1.1.5}$$

Then there exists an orthonormal sequence of continuous eigenfunctions, $\Phi_1, \Phi_2, \ldots$ in $\mathcal{L}_2[\mathcal{T}]$ and eigenvalues $\lambda_1 \geq \lambda_2 \geq \ldots \geq 0$, with

$$\int_{\mathcal{T}} R(s,t)\Phi_\nu(t)\,dt = \lambda_\nu \Phi_\nu(s), \quad \nu = 1, 2, \ldots, \tag{1.1.6}$$

$$R(s,t) = \sum_{\nu=1}^{\infty} \lambda_\nu \Phi_\nu(s)\Phi_\nu(t), \tag{1.1.7}$$

$$\int_{\mathcal{T}}\int_{\mathcal{T}} R^2(s,t)\,ds\,dt = \sum_{\nu=1}^{\infty} \lambda_\nu^2 < \infty. \tag{1.1.8}$$

See the Mercer–Hilbert–Schmidt theorems of Riesz and Sz.-Nagy (1955, pp. 242–246) for proofs of (1.1.6)–(1.1.8). Note that if we rewrite this result for the case $\mathcal{T} = (1, 2, \ldots, N)$, then (1.1.6)–(1.1.8) become

$$\begin{aligned} R\Phi_\nu &= \lambda_\nu \Phi_\nu, \\ R &= \Gamma D \Gamma', \\ \text{trace } R^2 &= \sum_{\nu=1}^{N} \lambda_\nu^2, \end{aligned}$$

where R is the $N \times N$ matrix with ijth entry $R(i,j)$, Φ_ν is the vector with jth entry $\Phi_\nu(j)$, D is the diagonal matrix with $\nu\nu$th entry λ_ν, and Γ is the orthogonal matrix with νth column Φ_ν.

We have the following lemma.

LEMMA 1.1.1. *Suppose* (1.1.5) *holds. If we let*

$$f_\nu = \int_T f(t)\Phi_\nu(t)\,dt, \tag{1.1.9}$$

then $f \in \mathcal{H}_R$ *if and only if*

$$\sum_{\nu=1}^{\infty} \frac{f_\nu^2}{\lambda_\nu} < \infty \tag{1.1.10}$$

and

$$\|f\|_R^2 = \sum_{\nu=1}^{\infty} \frac{f_\nu^2}{\lambda_\nu}. \tag{1.1.11}$$

Proof. The collection of all functions f with $\Sigma(f_\nu^2/\lambda_\nu) < \infty$ is clearly a Hilbert space with $\|f\|^2 = \Sigma(f_\nu^2/\lambda_\nu)$. We must show that R with

$$R(s,t) = \Sigma\lambda_\nu\Phi_\nu(s)\Phi_\nu(t)$$

is its r.k. That is, we must show that $R_t \in \mathcal{H}_R$ and

$$< f, R_t >= f(t), \quad f \in \mathcal{H}_R\,, \quad t \in \mathcal{T}$$

for $R_t(s) = R(t,s)$. Expanding f and $R(t,\cdot)$ in Fourier series with respect to $\Phi_1, \Phi_2, \ldots$, we have

$$\begin{aligned} f(\cdot) &\sim \sum_\nu f_\nu\Phi_\nu(\cdot), \\ R(t,\cdot) &\sim \sum_\nu \{\lambda_\nu\Phi_\nu(t)\}\Phi_\nu(\cdot)\,. \end{aligned}$$

$R_t \in \mathcal{H}_R$ since $\sum_\nu\{\lambda_\nu\Phi_\nu(t)\}^2/\lambda_\nu = \sum_\nu \lambda_\nu\Phi_\nu^2(t) = R(t,t) < \infty$ and

$$< f, R_t >=< f, R(t,\cdot) >= \sum_\nu f_\nu\{\lambda_\nu\Phi_\nu(t)\}/\lambda_\nu, \qquad t \in \mathcal{T}$$

using the inner product induced by the norm in (1.1.11). But

$$\sum_\nu f_\nu\{\lambda_\nu\Phi_\nu(t)\}/\lambda_\nu = \sum_\nu f_\nu\Phi_\nu(t) = f(t),$$

and the result is proved.

We remark that if we begin with R satisfying (1.1.5) and construct the Hilbert space of functions with $\sum(f_\nu^2/\lambda_\nu) < \infty$, it is easy to show that the evaluation functionals are bounded:

$$\begin{aligned} f(t) = \sum_{\nu=1}^{\infty} \frac{f_\nu\{\sqrt{\lambda_\nu}\Phi_\nu(t)\}}{\sqrt{\lambda_\nu}} &\leq \sqrt{\sum_{\nu=1}^{\infty}\frac{f_\nu^2}{\lambda_\nu}\sum_{\nu=1}^{\infty}\lambda_\nu\Phi_\nu^2(t)} \\ &= \|f\|\sqrt{R(t,t)} = \|f\|\;\|R_t\|. \end{aligned}$$

We remind the reader of the Karhunen–Loeve expansion. Suppose R is a covariance for which (1.1.5) holds, and let $X(t),\ t \in \mathcal{T}$ be a family of zero-mean Gaussian random variables with $EX(s)X(t) = R(s,t) = \sum_{\nu=1}^{\infty} \lambda_\nu \Phi_\nu(s)\Phi_\nu(t)$. Then $X(t), t \in \mathcal{T}$ has a (quadratic mean) representation

$$X(t) \sim \sum_{\nu=1}^{\infty} X_\nu \Phi_\nu(t),$$

where $X_1, X_2, \ldots$ are independent, Gaussian random variables with

$$EX_\nu = 0, \qquad EX_\nu^2 = \lambda_\nu$$

and

$$X_\nu = \int_{\mathcal{T}} X(s)\Phi_\nu(s)\, ds. \tag{1.1.12}$$

The integral in (1.1.12) is well defined in quadratic mean (see Cramer and Leadbetter (1967)). However, sample functions of $X(t),\ t \in \mathcal{T}$ are not (with probability 1) in $\mathcal{H}_R$, if R has more than a finite number of nonzero eigenvalues. We do not prove this (see Hajek (1962a)), but merely consider the following suggestion of this fact. Let

$$X_K(t) = \sum_{\nu=1}^{K} X_\nu \Phi_\nu(t), \qquad t \in \mathcal{T},$$

then for each fixed t, $X_K(t)$ tends to $X(t)$ in quadratic mean, since

$$E|X_K(t) - X(t)|^2 = E|\sum_{K+1}^{\infty} X_\nu \Phi_\nu(t)|^2 = \sum_{K+1}^{\infty} \lambda_\nu \Phi_\nu^2(t) \to 0;$$

however,

$$E\|X_K(\cdot)\|^2 = E\sum_{\nu=1}^{K} \frac{X_\nu^2}{\lambda_\nu} = K \to \infty \text{ as } K \to \infty.$$

This very important fact, namely, that the assumptions that $f \in \mathcal{H}_R$ and f a sample function from a zero-mean Gaussian stochastic process are *not* the same, will have important consequences later.

1.2 Reproducing kernel spaces on $[0,1]$ with norms involving derivatives.

We remind the reader of Taylor's theorem with remainder: If f is a real- valued function on $[0,1]$ with $m-1$ continuous derivatives and $f^{(m)} \in \mathcal{L}_2[0,1]$, then we may write

$$f(t) = \left\{ \sum_{\nu=0}^{m-1} \frac{t^\nu}{\nu!} f^{(\nu)}(0) \right\} + \left\{ \int_0^1 \frac{(t-u)_+^{m-1}}{(m-1)!} f^{(m)}(u)\, du \right\}, \tag{1.2.1}$$

where $(x)_+ = x$ for $x \geq 0$ and $(x)_+ = 0$ otherwise. Let $\mathcal{B}_m$ denote the class of functions satisfying the boundary conditions $f^{(\nu)}(0) = 0$, $\nu = 0, 1 \ldots, m-1$. If $f \in \mathcal{B}_m$ then

$$\begin{aligned} f(t) &= \int_0^1 \frac{(t-u)_+^{m-1}}{(m-1)!} f^{(m)}(u)\, du \\ &= \int_0^1 G_m(t,u) f^{(m)}(u)\, du, \quad \text{say}, \end{aligned} \tag{1.2.2}$$

where

$$G_m(t,u) = (t-u)_+^{m-1}/(m-1)!. \tag{1.2.3}$$

G_m is the *Green's function* for the problem $D^m f = g$, $f \in \mathcal{B}_m$, where D^m denotes the mth derivative. Equation (1.2.2) can be verified by interchanging the order of integration in $f(t) = \int_0^t dt_{m-1} \int_0^{t_{m-1}} dt_{m-2} \cdots \int_0^{t_1} f^{(m)}(u)\, du$. Denote by W_m^0 the collection of functions on $[0,1]$ with

$$\{f : f \in \mathcal{B}_m, f, f', \ldots, f^{(m-1)} \text{ absolutely continuous}, f^{(m)} \in \mathcal{L}_2\}.$$

It is not hard to show that W_m^0 is a Hilbert space with square norm $\|f\|^2 = \int_0^1 (f^{(m)}(t))^2\, dt$. We claim that W_m^0 is an r.k.h.s. with r.k.

$$R(s,t) = \int_0^1 G_m(t,u) G_m(s,u)\, du. \tag{1.2.4}$$

To show that the evaluation functionals are bounded, note that for $f \in W_m^0$ we have

$$f(s) = \int_0^1 G_m(s,u) f^{(m)}(u)\, du \tag{1.2.5}$$

so that by the Cauchy–Schwarz inequality

$$\begin{aligned} |f(s)| &\leq \sqrt{\int_0^1 G_m^2(s,u)\, du} \cdot \sqrt{\int_0^1 (f^{(m)}(u))^2\, du} \\ &= \sqrt{R\,(s,s)}\, \|f\|. \end{aligned}$$

To show that $R(\cdot,\cdot)$ is the r.k. for W_m^0 we must show that $R_t(\cdot) = R(t,\cdot)$ is in W_m^0 and that $< R_t, f >= f(t)$, all $f \in W_m^o$. But

$$R_t(v) = \int_0^1 G_m(v,u) G_m(t,u)\, du$$

and hence R_t is in W_m^0,

$$\left(\frac{\partial^m}{\partial v^m} R_t\right)(v) = G_m(t,v),$$

as can be seen by letting $f = R_t$ in (1.2.5). Thus

$$< f, R_t >=< R_t, f >= \int_0^1 G_m(t, v) f^{(m)}(v)\, dv = f(t).$$

Now let $\phi_\nu(t) = t^{\nu-1}/(\nu-1)!$ for $\nu = 1, 2, \ldots, m$ and denote the m-dimensional space of polynomials of degree $m-1$ or less spanned by $\phi_1, \cdots \phi_m$ as $\mathcal{H}_0$. Note that $D^m(\mathcal{H}_0) = 0$. Since

$$\begin{aligned} (D^{\mu-1}\phi_\nu)(0) &= 1, \quad \mu = \nu \\ &= 0, \quad \mu \neq \nu, \quad \mu, \nu = 1, \cdots, m, \end{aligned}$$

$\mathcal{H}_0$ endowed with the squared norm

$$\|\phi\|^2 = \sum_{\nu=0}^{m-1} [(D^\nu \phi)(0)]^2,$$

is an m-dimensional Hilbert space with $\phi_1, \ldots, \phi_m$ as an orthonormal basis, and it is not hard to show that then the r.k. for $\mathcal{H}_0$ is

$$\sum_{\nu=1}^{m} \phi_\nu(s)\phi_\nu(t).$$

To see this, let $R_t(\cdot) = \sum_{\nu=1}^m \phi_\nu(t)\phi_\nu(\cdot)$; then

$$< R_t, \phi_\alpha >= \sum_{\nu=1}^{m} \phi_\nu(t) < \phi_\nu, \phi_\alpha >= \phi_\alpha(t), \quad \alpha = 1, 2, \ldots, m.$$

We are now ready to construct the so-called Sobolev–Hilbert space W_m;

$$W_m : W_m[0,1] = \{f : \ f, f', \ldots, f^{m-1} \text{absolutely continuous}, f^{(m)} \in \mathcal{L}_2\}.$$

There are a number of ways to construct a norm on W_m. The norm we give here is given in Kimeldorf and Wahba (1971) and has associated with it an r.k. that will be particularly useful for our purposes. Different (but topologically equivalent) norms on this space will be introduced below and in Section 10.2. "Sobolev space" is the general term given for a function space (not necessarily a Hilbert space) whose norm involves derivatives. For more on Sobolev spaces, see Adams (1975). Each element in W_m has a Taylor series expansion (1.2.1) to order m and hence a unique decomposition

$$f = f_0 + f_1$$

with $f_0 \in \mathcal{H}_0$ and $f_1 \in W_m^o$, given by the first and second terms in brackets in (1.2.1). Furthermore, $\int_0^1 ((D^m f_0)(u))^2 du = 0$ and $\sum_{\nu=0}^{m-1} [(D^\nu f_1)(0)]^2 = 0$. Thus, denoting W_m^0 by $\mathcal{H}_1$, we claim

$$W_m = \mathcal{H}_0 \oplus \mathcal{H}_1,$$

and, if we endow W_m with the square norm

$$\|f\|^2 = \sum_{\nu=0}^{m-1} [(D^\nu f)(0)]^2 + \int_0^1 (D^m f)^2(u)\, du$$

then $\mathcal{H}_0$ and $\mathcal{H}_1$ will be perpendicular. With this norm, it is not hard to show that the r.k. for W_m is

$$R(s,t) = \sum_{\nu=1}^{m} \phi_\nu(s)\phi_\nu(t) + \int_0^1 G_m(s,u)G_m(t,u)\, du$$

where G_m is given by (1.2.3). The reproducing kernel for the direct sum of two perpendicular subspaces is the sum of the r.k.'s (see Aronszajn (1950)). An important geometrical fact that we will use later is that the penalty functional $J_m(f) = \int_0^1 (f^{(m)}(u))^2\, du$ may be written

$$J_m(f) = \|P_1 f\|^2_{W_m}$$

where P_1 is the orthogonal projection of f onto $\mathcal{H}_1$ in W_m.

We may replace D^m in $\int_0^1 (D^m f)^2\, du$ by more general differential operators. Let $a_1, a_2, \ldots, a_m$ be strictly positive functions with $a_i(0) = 1$ and as many derivatives as needed and let

$$L_m = D\frac{1}{a_1}D\frac{1}{a_2}\cdots D\frac{1}{a_m}.$$

Also let

$$\begin{aligned} M_0 &= I \text{ (the identity)} \\ M_1 &= D\frac{1}{a_m} \\ M_2 &= D\frac{1}{a_{m-1}}\,D\frac{1}{a_m} \\ &\vdots \\ M_{m-1} &= D\frac{1}{a_2}\,D\frac{1}{a_3}\cdots D\frac{1}{a_m} \end{aligned}$$

and let $\omega_1, \ldots, \omega_m$ be defined by

$$\begin{aligned} \omega_1(t) &= a_m(t) \\ \omega_2(t) &= a_m(t)\int_0^t a_{m-1}(t_{m-1})\, dt_{m-1} \\ &\vdots \\ \omega_m(t) &= a_m(t)\int_0^t a_{m-1}(t_{m-1})\, dt_{m-1} \\ &\quad \cdot \int_0^{t_{m-1}} a_{m-2}(t_{m-2})\, dt_{m-2} \cdots \int_0^{t_2} a_1(t_1)\, dt_1. \end{aligned}$$

Note that

$$
\begin{aligned}
(M_{\mu-1}\omega_\nu)(0) &= 1, \qquad \mu=\nu \\
&= 0, \qquad \mu\neq\nu, \qquad \mu,\nu=1,\ldots,m.
\end{aligned}
$$

The $\{\omega_\nu\}$ are an "extended Tchebycheff system" and share the following property with the polynomials $\phi_1,\ldots,\phi_m$. Let $t_1,\ldots,t_n$ be distinct, with $n\geq m$; then the $n\times m$ matrix T with i,νth entry $\omega_\nu(t_i)$ is of rank m (see Karlin (1968)).

Now, let $\tilde{\mathcal{B}}_m$ denote the class of functions satisfying the boundary conditions

$$(M_\nu f)(0)=0, \qquad \nu=0,1,\ldots,m-1,$$

and let $\tilde{G}_m$ be the Green's function for the problem $L_m f=g,\ f\in\tilde{\mathcal{B}}_m$. We have $f\in\tilde{\mathcal{B}}_m\Rightarrow$

$$
\begin{aligned}
f(t) &= a_m(t)\int_0^t a_{m-1}(t_{m-1})\,dt_{m-1} \\
&\quad\cdot\int_0^{t_{m-1}} a_{m-2}(t_{m-2})\,dt_{m-2}\cdots\int_0^{t_1}(L_m f)(u)\,du \\
&= \int_0^t (L_m f)(u)\,du\{a_m(t)\int_u^t a_1(t_1)\,dt_1 \qquad (1.2.6)\\
&\quad\cdot\int_{t_1}^t a_2(t_2)\,dt_2\cdots\int_{t_{m-2}}^t a_{m-1}(t_{m-1})\,dt_{m-1}\} \\
&= \int_0^t \tilde{G}_m(t,u)(L_m f)(u)\,du,
\end{aligned}
$$

where $\tilde{G}_m(t,m)$ is equal to the expression in brackets in (1.2.6).

Let $\tilde{W}_m^0$ be the collection of functions on $[0,1]$ given by

$$\{f:\ f\in\tilde{\mathcal{B}}_m,\ M_0f,\ M_1f,\cdots,M_{m-1}f \text{ absolutely continuous, } L_mf\in\mathcal{L}_2\}.$$

Then by the same arguments as before, $\tilde{W}_m^0$ is an r.k.h.s. with the squared norm $\|f\|^2=\int_0^1(L_mf)^2(u)\,du$, and reproducing kernel

$$R(s,t)=\int_0^1 \tilde{G}_m(s,u)\tilde{G}_m(t,u)\,du.$$

Letting $\mathcal{H}_0$ be span $\{\omega_1,\ldots,\omega_m\}$ and $\mathcal{H}_1$ be $\tilde{W}_m^0$, then letting $\tilde{W}_m$ be the Hilbert space

$$\tilde{W}_m=\mathcal{H}_0\oplus\mathcal{H}_1,$$

we have that $\tilde{W}_m$ is an r.k.h.s. with

$$\|f\|^2=\sum_{\nu=0}^{m-1}[(M_\nu f)(0)]^2+\int_0^1(L_mf)^2\,du$$

and r.k.

$$\sum_{\nu=0}^{m-1} \omega_\nu(s)\omega_\nu(t) + \int_0^1 \tilde{G}_m(s,u)\tilde{G}_m(t,u)\,du.$$

Furthermore, we have the geometrical relation

$$\int_0^1 (L_m f)^2\,du = \|P_1 f\|^2_{\tilde{W}_m},$$

where P_1 is the orthogonal projection in $\tilde{W}_m$ onto $\mathcal{H}_1$.

We have, for $f \in \tilde{W}_m$, the *generalized Taylor series expansion*

$$f(t) = \sum_{\nu=1}^{m} \omega_\nu(t)(M_{\nu-1}f)(0) + \int_0^t \tilde{G}_m(t,u)(L_m f)(u)\,du.$$

We remark that W_m and $\tilde{W}_m$ are topologically equivalent; they have the same Cauchy sequences. Another topologically equivalent norm involving boundary rather than initial values will be introduced in Section 10.2.

1.3 The special and general spline smoothing problems.

The data model associated with the special spline smoothing problem is

$$y_i = f(t_i) + \epsilon_i, \qquad i = 1, 2, \ldots, n \tag{1.3.1}$$

where $t \in \mathcal{T} = [0,1]$, $f \in W_m$, and $\epsilon = (\epsilon_1, \ldots, \epsilon_n)' \sim \mathcal{N}(0, \sigma^2 I)$. An estimate of f is obtained by finding $f \in W_m$ to minimize

$$\frac{1}{n}\sum_{i=1}^{n}(y_i - f(t_i))^2 + \lambda \int_0^1 (f^{(m)}(u))^2\,du. \tag{1.3.2}$$

The data model associated with the general spline smoothing problem is

$$y_i = L_i f + \epsilon_i, \quad i = 1, 2, \ldots, n \tag{1.3.3}$$

where ϵ is as before. Now $\mathcal{T}$ is arbitrary, $f \in \mathcal{H}_R$, a given r.k.h.s. of functions on $\mathcal{T}$, and $L_1, \ldots, L_n$ are bounded linear functionals on $\mathcal{H}_R$. $\mathcal{H}_R$ is supposed to have a decomposition

$$\mathcal{H}_R = \mathcal{H}_0 \oplus \mathcal{H}_1$$

where $\dim \mathcal{H}_0 = M \le n$. An estimate of f is obtained by finding $f \in \mathcal{H}_R$ to minimize

$$\frac{1}{n}\sum_{i=1}^{n}(y_i - L_i f)^2 + \lambda \|P_1 f\|^2_R, \tag{1.3.4}$$

where P_1 is the orthogonal projection of f onto $\mathcal{H}_1$, in $\mathcal{H}_R$.

One of the useful properties of reproducing kernels is that from them one can obtain the representer of any bounded linear functional. Let η_i be the representer for L_i, that is,

$$<\eta_i, f> = L_i f, \quad f \in \mathcal{H}_R.$$

Then

$$\eta_i(s) = < \eta_i, R_s > = L_i R_s = L_{i(\cdot)} R(s, \cdot) \tag{1.3.5}$$

where $L_{i(\cdot)}$ means L_i is applied to what follows as a function of $(\cdot)$. That is, one can apply L_i to $R(s,t)$ considered as a function of t, to obtain $\eta_i(s)$. For example, if $L_i f = \int w_i(u) f(u)\, du$, then $\eta_i(s) = \int w_i(u) R(s,u)\, du$, and if $L_i f = f'(t_i)$, then $\eta_i(s) = (\partial/\partial u) R(s,u)|_{u=t_i}$. On the other hand L_i is a bounded linear functional on $\mathcal{H}_R$ only if $\eta_i(\cdot)$ obtained by $\eta_i(s) = L_{i(\cdot)} R(s,\cdot)$ is a well-defined element of $\mathcal{H}_R$. To see the argument behind this note that if $L_i f = \sum_\ell a_\ell f(t_\ell)$ for any finite sum, then its representer is $\eta_i = \sum_\ell a_\ell R_{t_\ell}$, and any η_i in $\mathcal{H}_R$ will be a limit of sums of this form and will be the representer of the limiting bounded linear functional. As an example, $L_i f = f'(t_i)$ a bounded linear functional in $\mathcal{H}_R \Rightarrow \eta_i = \lim_{h \to 0} (1/h)(R_{t_i + h} - R_{t_i})$, where the limit is in the norm topology, which then entails that

$$\frac{\partial}{\partial t} R(t,s)|_{t=t_i} = \eta_i(s) \quad \text{with } \eta_i \in \mathcal{H}_R.$$

$f^{(k)}(t_i)$ can be shown to be a bounded linear functional in W_m for $k = 0, 1, \ldots, m-1$. More details can be found in Aronszajn (1950).

We will now find an explicit formula for the minimizer of (1.3.4), which can now be written

$$\frac{1}{n} \sum_{i=1}^{n} (y_i - < \eta_i, f >)^2 + \lambda \, \|P_1 f\|_R^2. \tag{1.3.6}$$

THEOREM 1.3.1. *Let $\phi_1, \ldots, \phi_M$ span the null space $(\mathcal{H}_0)$ of P_1 and let the $n \times M$ matrix $T_{n \times M}$ defined by*

$$T_{n \times M} = \{L_i \phi_\nu\}_{i=1}^{n} {}_{\nu=1}^{M} \tag{1.3.7}$$

be of full column rank. Then f_λ, the minimizer of (1.3.6), *is given by*

$$f_\lambda = \sum_{\nu=1}^{M} d_\nu \phi_\nu + \sum_{i=1}^{n} c_i \xi_i \tag{1.3.8}$$

where

$$\begin{aligned} \xi_i &= P_1 \eta_i, \\ d = (d_1, \ldots, d_M)' &= (T'M^{-1}T)^{-1} T'M^{-1} y, \\ c = (c_1, \ldots, c_n)' &= M^{-1}(I - T(T'M^{-1}T)^{-1}T'M^{-1}) y, \\ M &= \Sigma + n\lambda I, \\ \Sigma &= \{< \xi_i, \xi_j >\}. \end{aligned} \tag{1.3.9}$$

(Do not confuse the index M and the matrix M.)

Before giving the proof we make a few remarks concerning the ingredients of the minimizer. Letting $\mathcal{H}_R = \mathcal{H}_0 \oplus \mathcal{H}_1$, with $\mathcal{H}_0 \perp \mathcal{H}_1$, where

$$R(s,t) = R^0(s,t) + R^1(s,t)$$

and R^α is the r.k. for $\mathcal{H}_\alpha$, $\alpha = 0, 1$, we then have

$$\begin{aligned} \xi_i(t) = <\xi_i, R_t> &= <P_1\eta_i, R_t> = <\eta_i, P_1 R_t> \\ &= <\eta_i, R_t^1> \\ &= L_i R_t^1 \end{aligned} \quad (1.3.10)$$

where R_t^1 is the representer of evaluation at t in $\mathcal{H}_1$. We have used that the projection P_1 is self-adjoint. Furthermore,

$$<\xi_i, \xi_j> = <\eta_i, \xi_j>$$

since $<\eta_i - \xi_i, \xi_j> = 0$, so that

$$<\xi_i, \xi_j> = L_i \xi_j = L_{i(s)} L_{j(t)} R^1(s, t).$$

To prove the theorem, let the minimizer f_λ be of the form

$$f_\lambda = \sum_{\nu=1}^{M} d_\nu \phi_\nu + \sum_{i=1}^{n} c_i \xi_i + \rho \quad (1.3.11)$$

where ρ is some element in $\mathcal{H}_R$ perpendicular to $\phi_1, \ldots, \phi_M$, $\xi_1, \ldots, \xi_n$. Any element in $\mathcal{H}_R$ has such a representation by the property of Hilbert spaces. Then (1.3.6) becomes

$$\frac{1}{n}\|y - (\Sigma c + Td)\|^2 + \lambda(c'\Sigma c + \|\rho\|^2) \quad (1.3.12)$$

and we must find c, d, and ρ to minimize this expression. It is then obvious that $\|\rho\|^2 = 0$, and a straightforward calculation shows that the minimizing c and d of

$$\frac{1}{n}\|y - (\Sigma c + Td)\|^2 + \lambda c'\Sigma c \quad (1.3.13)$$

are given by

$$d = (T'M^{-1}T)^{-1}T'M^{-1}y, \quad (1.3.14)$$
$$c = M^{-1}(I - T(T'M^{-1}T)^{-1}T'M^{-1})y. \quad (1.3.15)$$

These formulae are quite unsuitable for numerical work, and, in fact, were quite impractical when they appeared in Kimeldorf and Wahba (1971). Utreras (1979) provided an equivalent set of equations with more favorable properties, and another improvement was given in Wahba (1978b) with the aid of an anonymous referee, who was later unmasked as Silverman. Multiplying the left and right sides of (1.3.15) by M and substituting in (1.3.14) gives (1.3.16) and multiplying (1.3.15) by T' gives (1.3.17):

$$Mc + Td = y, \quad (1.3.16)$$
$$T'c = 0, \quad (1.3.17)$$

these being equivalent to (1.3.14) and (1.3.15).

To compute c and d, let the QR decomposition (see Dongarra et al. (1979)) of T be

$$T = (Q_1 : Q_2) \begin{pmatrix} R \\ 0 \end{pmatrix} \tag{1.3.18}$$

where Q_1 is $n \times M$ and Q_2 is $n \times (n - M)$, $Q = (Q_1 : Q_2)$ is orthogonal and R is upper triangular, with $T'Q_2 = 0_{M\times(n-M)}$. Since $T'c = 0$, c must be in the column space of Q_2, giving $c = Q_2\gamma$ for some γ an $n - M$ vector. Substituting $c = Q_2\gamma$ into (1.3.16) and multiplying through by Q_2', recalling that $Q_2'T = 0$, gives

$$\begin{aligned} Q_2'MQ_2\gamma &= Q_2'y, \\ c &= Q_2\gamma = Q_2(Q_2'MQ_2)^{-1}Q_2'y, \end{aligned} \tag{1.3.19}$$

and multiplying (1.3.16) by Q_1' gives

$$Rd = Q_1'(y - Mc). \tag{1.3.20}$$

For later use the influence matrix $A(\lambda)$ will play an important role. $A(\lambda)$ is defined as the matrix satisfying

$$\begin{pmatrix} L_1 f_\lambda \\ \vdots \\ L_n f_\lambda \end{pmatrix} = A(\lambda)y. \tag{1.3.21}$$

To obtain a simple formula for $I - A(\lambda)$ we observe by substitution in (1.3.11) with $\rho = 0$ that

$$\begin{pmatrix} L_1 f_\lambda \\ \vdots \\ L_n f_\lambda \end{pmatrix} = Td + \Sigma c. \tag{1.3.22}$$

Subtracting this from (1.3.16) gives

$$(I - A(\lambda))y = n\lambda c = n\lambda Q_2(Q_2'MQ_2)^{-1}Q_2'y$$

for any y, thus

$$I - A(\lambda) = n\lambda Q_2(Q_2'MQ_2)^{-1}Q_2'. \tag{1.3.23}$$

Of course Q_2 may be replaced by any $n \times (n - M)$ matrix whose columns are any orthonormal set perpendicular to the M columns of T. We will discuss efficient numerical methods for computing c and d in conjunction with data-based methods for choosing λ later.

For the special spline smoothing problem we will demonstrate that f_λ of (1.3.8) is a natural polynomial spline. Here

$$\begin{aligned} L_i f &= f(t_i), \\ \|P_1 f\|^2 &= \int_0^1 (f^{(m)}(u))^2\,du, \\ R^1(s,t) &= \int_0^1 \frac{(s-u)_+^{m-1}(t-u)_+^{m-1}}{[(m-1)!]^2}\,du \end{aligned}$$

and

$$\xi_i(\cdot) = R^1(\cdot, t_i).$$

It is easy to check that here

$$\begin{aligned} \xi_i(\cdot) &\in \pi^{2m-1} \text{ for } s \in [0, t_i] \\ &\in \pi^{m-1} \text{ for } s \in [t_i, 1], \end{aligned}$$

and

$$\xi_i(\cdot) \in C^{2m-2}.$$

Thus

$$\begin{aligned} f_\lambda(t) = \sum_{\nu=1}^{m} d_\nu \phi_\nu(t) + \sum_{i=1}^{n} c_i \xi_i(t) &\in \pi^{m-1} \text{ for } t \in [t_n, 1] \\ &\in \pi^{2m-1} \text{ for } t \in [t_i, t_{i+1}] \\ &\in C^{2m-2}. \end{aligned}$$

We will show that the condition $T'c = 0$ guarantees that $f_\lambda \in \pi^{m-1}$ for $t \in [0, t_1]$, as follows. For $t < t_1$, we can remove the "+" in the formula for ξ_i and write

$$\sum_{i=1}^{n} c_i \xi_i(t) = \int_0^t \frac{(t-u)^{m-1}}{(m-1)!} \sum_{i=1}^{n} c_i (t_i - u)^{m-1} du, \ t < t_1. \tag{1.3.24}$$

But $\sum_{i=1}^{n} c_i t_i^k = 0$ for $k = 0, 1, \ldots, m-1$ since $T'c = 0$, so that (1.3.24) is 0 for $t < t_1$ and the result is proved.

We remark that it can be shown that $\lim_{\lambda \to \infty} f_\lambda$ is the least squares regression onto $\phi_1, \ldots, \phi_M$ and $\lim_{\lambda \to 0} f_\lambda$ is the interpolant to $L_i f = y_i$ in $\mathcal{H}$ that minimizes $\|P_1 f\|$. The important choice of λ from the data will be discussed later.

1.4 The duality between r.k.h.s. and stochastic processes.

Later we will show how spline estimates are also Bayes estimates, with a certain prior on f. This is no coincidence, but is a consequence of the duality between the Hilbert space spanned by a family of random variables and its associated r.k.h.s. The discussion of this duality follows Parzen (1962, 1970).

Let $X(t)$, $t \in \mathcal{T}$ be a family of zero-mean Gaussian random variables with $EX(s)X(t) = R(s,t)$. Let $\mathcal{X}$ be the *Hilbert space spanned by* $X(t)$, $t \in \mathcal{T}$. This is the collection of all random variables of the form

$$Z = \Sigma a_j X(t_j) \tag{1.4.1}$$

$t_j \in \mathcal{T}$, with inner product $< Z_1, Z_2 >= EZ_1 Z_2$, and all of their quadratic mean limits, i.e. Z is in $\mathcal{X}$ if and only if there is a sequence Z_l, $l = 1, 2, \ldots$ of random variables each of the form (1.4.1), with $\lim_{l \to \infty} E(Z - Z_l)^2 =$

$\|Z - Z_l\|^2 \to 0$. Letting $\mathcal{H}_R$ be the r.k.h.s. with r.k. R, we will see that $\mathcal{H}_R$ is isometrically isomorphic to $\mathcal{X}$, that is, there exists a 1:1 inner product preserving correspondence between the two spaces. The correspondence is given by Table 1.1. This correspondence is clearly 1:1 and preserves inner products,

TABLE 1.1
The 1:1 correspondence between $\mathcal{H}_R$ and $\mathcal{X}$.

$\mathcal{X}$		$\mathcal{H}_R$
$X(t)$	$\sim$	R_t
$\Sigma a_j X(t_j)$	$\sim$	$\Sigma a_j R_{t_j}$
$\lim \Sigma a_j X(t_j)$	$\sim$	$\lim \Sigma a_j R_{t_j}$.

since

$$< X(s), X(t) >= EX(s)X(t) = R(s,t) =< R_s, R_t > .$$

Let L be a bounded linear functional in $\mathcal{H}_R$ with representer η. Then η is the limit of a sequence of elements of the form $\Sigma a_{t_l} R_{t_l}$, by construction of $\mathcal{H}_R$. The random variable Z corresponds to η if Z is the (quadratic mean) limit of the corresponding sequence of random variables $\Sigma a_{t_l} X(t_l)$ and we can finally denote this limiting random variable by LX (although $X \notin \mathcal{H}_R$ and we do not think of L as a bounded linear functional applied to X). Then $EZX(t) = < \eta, R_t >= \eta(t) = LR_t$. Examples are $Z = \int w(t)X(t)\,dt$ and $Z = X'(t)$, if they exist.

We are now ready to give a simple example of the duality between Bayesian estimation on a family of random variables and optimization in an r.k.h.s. Consider $X(t)$, $t \in \mathcal{T}$ a zero-mean Gaussian stochastic process with $EX(s)X(t) = R(s,t)$. Fix t for the moment and compute $E\{X(t)|X(t_1) = x_1, \ldots, X(t_n) = x_n\}$. The joint covariance matrix of $X(t)$, $X(t_1), \ldots, X(t_n)$ is

$$\begin{pmatrix} R(t,t) & R(t,t_1), \ldots, & R(t,t_n) \\ R(t,t_1) & & \\ \vdots & R_n & \\ R(t,t_n) & & \end{pmatrix}$$

where R_n is the $n \times n$ matrix with ijth entry $R(t_i, t_j)$. We will assume for simplicity in this example that R_n is strictly positive definite. Using properties of the multivariate normal distribution, as given, e.g., in Anderson (1958), we have

$$\begin{aligned} & E\{X(t) \mid X(t_i) = x_i,\ i = 1, \ldots, n\} \\ & = (R(t,t_1), \ldots, R(t,t_n))\, R_n^{-1} x = \hat{f}(t), \end{aligned} \tag{1.4.2}$$

say. The Gaussianness is not actually being used, except to call $\hat{f}(t)$ a conditional expectation. If $\hat{f}(t)$ were just required to be the minimum variance unbiased linear estimate of $X(t)$, given the data, the result $\hat{f}(t)$ would be the same, independent of the form of the joint distribution and depending only on the first and second moments.

Now consider the following problem. Find $f \in \mathcal{H}_R$, the r.k.h.s. with r.k. R, to minimize $\|f\|^2$ subject to $f(t_i) = x_i$, $i = 1, \ldots, n$. By a special case of the argument given before, f must be of the form

$$f = \sum_{j=1}^{n} c_j \, R_{t_j} + \rho$$

for some $\rho \perp$ to $R_{t_1}, \ldots, R_{t_n}$, that is, ρ satisfies $< R_{t_i}, \rho >= \rho(t_i) = 0$, $i = 1, \ldots, n$. $\|f\|^2 = c' R_n c + \|\rho\|^2$ and so $\|\rho\| = 0$. Setting $f(t_i) = \sum_{j=1}^{n} c_j R_{t_j}(t_i) = x_i$, $i = 1, \ldots, n$ gives $x = (x_1, \ldots, x_n)' = R_n c$, and so the minimizer is f given by

$$f = x' \, R_n^{-1} \begin{pmatrix} R_{t_1} \\ \vdots \\ R_{t_n} \end{pmatrix} = \hat{f},$$

which is exactly equal to $\hat{f}$ of (1.4.2)!

1.5 The smoothing spline and the generalized smoothing spline as Bayes estimates.

We first consider

$$W_m = \mathcal{H}_0 \oplus W_m^0,$$

where W_m^0 has the r.k.

$$R^1(s,t) = \int_0^1 \frac{(s-u)_+^{m-1}}{(m-1)!} \, \frac{(t-u)_+^{m-1}}{(m-1)!} \, du. \tag{1.5.1}$$

Let

$$X(t) = \int_0^1 \frac{(t-u)_+^{m-1}}{(m-1)!} \, dW(u) \tag{1.5.2}$$

where $W(\cdot)$ is the Wiener process. Formally, $X \subset \mathcal{B}_m$, $D^m X = dW =$ "white noise." $X(\cdot)$ is the $m-1$ fold integrated Wiener process described in Shepp (1966). We remind the reader that the Wiener process is a zero-mean Gaussian stochastic process with stationary independent increments and $W(0) = 0$. Stationary, independent increments means, for any s_1, s_2, s_3, s_4, the joint distribution of $W(s_2) - W(s_1)$ and $W(s_4) - W(s_3)$ is the same as that of $W(s_2 + h) - W(s_1 + h)$ and $W(s_4 + h) - W(s_3 + h)$, and, if the intervals $[s_1, s_2]$ and $[s_3, s_4]$ are nonoverlapping then $W(s_2) - W(s_1)$ and $W(s_4) - W(s_3)$ are independent. Integrals of the form

$$\int_0^1 g(u) \, dW(u) \tag{1.5.3}$$

are defined as quadratic-mean limits of the Riemann–Stieltjes sums

$$\sum g(u_l)\,[W(u_{l+1}) - W(u_l)] \tag{1.5.4}$$

for partitions $\{u_1, \ldots, u_L\}$ of $[0,1]$ (see Cramer and Leadbetter (1967, Chap. 5)). It can be shown to follow from the stationary independent increments property, that $E[W(u+h) - W(u)]^2 = \text{const. } h$, for some constant, which we will take here to be 1. Using the definition of (1.5.3) as a limit of the form (1.5.4), it can be shown using the independent increments property, that if g_1 and g_2 are in $\mathcal{L}_2[0,1]$, then

$$E\int_0^1 g_1(u)\,dW(u)\int_0^1 g_2(u)\,dW(u) = \int_0^1 g_1(u)g_2(u)\,du. \tag{1.5.5}$$

Thus,

$$EX(s)X(t) = \int_0^1 \frac{(t-u)_+^{m-1}}{(m-1)!}\,\frac{(s-u)_+^{m-1}}{(m-1)!}\,du = R^1(s,t) \tag{1.5.6}$$

and the Hilbert space spanned by the $m-1$ fold integrated Wiener process is isometrically isomorphic to W_m^0.

We will consider two types of Bayes estimates, both of which lead to a smoothing spline estimate. The first was given in Kimeldorf and Wahba (1971) and might be called the "fixed effects" model, and the second might be called the "random effects model with an improper prior," and was given in Wahba (1978b).

The first model is

$$\begin{aligned} F(t) &= \sum_{\nu=1}^{m} \theta_\nu \phi_\nu(t) + b^{1/2}X(t), \quad t \in [0,1], \\ Y_i &= F(t_i) + \epsilon_i, \quad i = 1, \ldots, n. \end{aligned} \tag{1.5.7}$$

Here $\theta = (\theta_1, \ldots, \theta_n)'$ is considered to be a fixed, but unknown, vector, b is a positive constant, $X(t)$, $t \in [0,1]$ is a zero-mean Gaussian stochastic process with covariance $R^1(s,t)$ of (1.5.6), and $\epsilon \sim \mathcal{N}(0, \sigma^2 I)$. We wish to construct an estimate of $F(t)$, $t \in \mathcal{T}$, given $Y_i = y_i$, $i = 1, \ldots, n$.

An estimate $\hat{F}(t)$ of $F(t)$ will be called unbiased with respect to θ if

$$E(\hat{F}(t)|\theta) = E(F(t)|\theta).$$

(Here, t is considered fixed.) Let $\hat{F}(t)$ be the minimum variance, unbiased (with respect to θ) linear estimate of $F(t)$ given $Y_i = y_i$, $i = 1, \ldots, n$. That is,

$$\hat{F}(t) = \sum_{j=1}^{n} \beta_j(t) y_j$$

for some $\beta_j(t)$ (linearity), and $\hat{F}(t)$ minimizes

$$E(\hat{F}(t) - F(t))^2$$

(minimum variance) subject to

$$E(\hat{F}(t) - F(t)|\theta) = 0 \quad \text{for all } t \in [0,1].$$

We have the following theorem.

THEOREM 1.5.1. *Let f_λ be the minimizer in W_m of*

$$\frac{1}{n}\sum_{i=1}^{n}(y_i - f(t_i))^2 + \lambda \int_0^1 (f^{(m)}(u))^2\, du.$$

Then

$$\hat{F}(t) = f_\lambda(t)$$

with $\lambda = \sigma^2/nb$.

Proof. A proof can be obtained by straightforward calculation (see Kimeldorf and Wahba (1971)).

The general version of this theorem follows. Let $\mathcal{H} = \mathcal{H}_0 \oplus \mathcal{H}_1$ where $\mathcal{H}_0$ is spanned by $\phi_1, \ldots, \phi_M$, and $\mathcal{H}_1$ has r.k. $R^1(s,t)$. Let

$$F(t) = \sum_{\nu=1}^{M} \theta_\nu \phi_\nu(t) + b^{1/2} X(t), \quad t \in \mathcal{T}$$

where θ is as before and $EX(s)X(t) = R^1(s,t)$. Let $L_1, \ldots, L_n$ be bounded linear functionals on $\mathcal{H}$; then $\sum_{\nu=1}^{M} \theta_\nu L_i \phi_\nu$ is a well-defined constant and $b^{1/2} L_i X$ is a well-defined random variable in the Hilbert space spanned by $X(t)$, $t \in \mathcal{T}$. Let

$$Y_i = L_i F + \epsilon_i, \quad i = 1, \ldots, n$$

where ϵ is as before. Here and elsewhere it is assumed that the $n \times M$ matrix T with $i\nu$th element $L_i\phi_\nu$ is of rank M (that is, least squares regression on $\phi_1, \ldots, \phi_M$ is uniquely defined). Let L_0 be another bounded linear functional on $\mathcal{H}$. The goal is to estimate $L_0 F$ (again a well-defined random variable), given $Y_i = y_i$, $i = 1, \ldots, n$. Call the estimate $\widehat{L_0 F}$. Let $\widehat{L_0 F}$ be the minimum variance, linear, unbiased with respect to θ estimate. Then

$$\widehat{L_0 F} = \sum_{j=1}^{n} \beta_j y_j$$

where $\beta = (\beta_1, \ldots, \beta_n)$ is chosen to minimize

$$E(\widehat{L_0 F} - L_0 F)^2$$

subject to

$$E[(\widehat{L_0 F} - L_0 F)|\theta] = 0.$$

We have the following theorem.

THEOREM 1.5.2.

$$\widehat{L_0 F} = L_0 f_\lambda$$

where f_λ is the minimizer in $\mathcal{H}$ of

$$\frac{1}{n}\sum_{i=1}^{n}(y_i - L_i f)^2 + \lambda\|P_1 f\|^2$$

with $\lambda = \sigma^2/nb$.

This theorem says that if you want to estimate $L_0 F$, then you can find the generalized smoothing spline f_λ for the data and take $L_0 f_\lambda$ as the estimate.

A practical application of this result that we will return to later is the estimation of $f'(t)$ given data:

$$y_i = f(t_i) + \epsilon_i, \quad i = 1, 2, \ldots, n.$$

One can take the smoothing spline for the data and use its derivative as an estimate of f'.

The second, or "random effects model with an improper prior," leads to the same smoothing spline, and goes as follows:

$$\begin{aligned} F(t) &= \sum_{\nu=1}^{M} \theta_\nu \phi_\nu(t) + b^{1/2} X(t), \\ Y_i &= L_i F + \epsilon_i, \end{aligned} \tag{1.5.8}$$

where everything is as before except θ, which is assumed to be $\mathcal{N}(0, aI)$, and we will let $a \to \infty$.

THEOREM 1.5.3. *Let*

$$\hat{F}_a(t) = E(F(t)|Y_i = y_i,\ i = 1, \ldots, n)$$

and let f_λ be the minimizer of

$$\frac{1}{n}\sum_{i=1}^{n}(y_i - L_i f)^2 + \lambda\|P_1 f\|^2$$

with $\lambda = \sigma^2/nb$. Then, for each fixed t,

$$\lim_{a\to\infty} \hat{F}_a(t) = f_\lambda(t).$$

To prove this, by the correspondence between $\mathcal{H}_{R^1}$ and the Hilbert space spanned by $X(t), t \in \mathcal{T}$, we have

$$\begin{aligned} E(L_i X)X(t) &= L_{i(s)} R^1(s,t) = \xi_i(t), \\ EL_i X L_j X &= L_{i(s)} L_{j(t)} R^1(s,t) = \langle \xi_i, \xi_j \rangle . \end{aligned}$$

Then, letting $Y = (Y_1, \ldots, Y_n)'$, we have

$$EF_a(t)Y = aT \begin{pmatrix} \phi_1(t) \\ \vdots \\ \phi_M(t) \end{pmatrix} + b \begin{pmatrix} \xi_1(t) \\ \vdots \\ \xi_n(t) \end{pmatrix},$$

$$EYY' = aTT' + b\Sigma + \sigma^2 I \tag{1.5.9}$$

where T and Σ are as in (1.3.7) and (1.3.9).

Setting $\lambda = \sigma^2/nb$, $\eta = a/b$ and $M = \Sigma + n\lambda I$ gives

$$\begin{aligned} E(F_a(t)|Y = y) &= (\phi_1(t), \ldots, \phi_M(t))\eta T'(\eta TT' + M)^{-1}y \\ &\quad + (\xi_1(t), \ldots, \xi_n(t))(\eta TT' + M)^{-1}y. \end{aligned} \tag{1.5.10}$$

Comparing (1.3.14), (1.3.15), and (1.5.8), it only remains to show that

$$\lim_{\eta \to \infty} \eta T'(\eta TT' + M)^{-1} = (T'M^{-1}T)^{-1}T'M^{-1} \tag{1.5.11}$$

and

$$\lim_{\eta \to \infty} (\eta TT' + M)^{-1} = M^{-1}(I - T(T'M^{-1}T)^{-1}T'M^{-1}). \tag{1.5.12}$$

It can be verified that

$$\begin{aligned} &(\eta TT' + M)^{-1} \\ &\quad = M^{-1} - M^{-1}T(T'M^{-1}T)^{-1}\left\{I + \eta^{-1}(T'M^{-1}T)^{-1}\right\}^{-1} T'M^{-1}, \end{aligned}$$

expanding in powers of η and letting $\eta \to \infty$ completes the proof.

CHAPTER 2

More Splines

2.1 Splines on the circle.

Splines on the circle can be obtained by imposing periodic boundary conditions on splines in W_m, but it is more instructive to describe splines on the circle from the beginning since the eigenfunctions and eigenvalues of the associated reproducing kernel have a particularly simple form.

Let W_m^0 (per) be the collection of all functions on $[0,1]$ of the form

$$f(t) \sim \sqrt{2}\sum_{\nu=1}^{\infty} a_\nu \cos 2\pi\nu t + \sqrt{2}\sum_{\nu=1}^{\infty} b_\nu \sin 2\pi\nu t$$

with

$$\sum_{\nu=1}^{\infty}(a_\nu^2 + b_\nu^2)(2\pi\nu)^{2m} < \infty. \tag{2.1.1}$$

Since

$$\frac{d^m}{dt^m}\left\{\begin{array}{c}\cos 2\pi\nu t,\\ \sin 2\pi\nu t\end{array}\right\} = (2\pi\nu)^m \times \left\{\begin{array}{c}\pm\sin 2\pi\nu t\\ \pm\cos 2\pi\nu t\end{array}\right\}, \tag{2.1.2}$$

then if (2.1.1) holds, we have

$$\sum_{v=1}^{\infty}(a_\nu^2 + b_\nu^2)(2\pi\nu)^{2m} = \int_0^1 (f^{(m)}(u))^2\, du. \tag{2.1.3}$$

It is easy to see that the r.k. $R^1(s,t)$ for W_m^0 (per) is

$$R^1(s,t) = \sum_{\nu=1}^{\infty}\frac{2}{(2\pi\nu)^{2m}}[\cos 2\pi\nu s \cos 2\pi\nu t + \sin 2\pi\nu s \sin 2\pi\nu t] \tag{2.1.4}$$

$$= \sum_{\nu=1}^{\infty}\frac{2}{(2\pi\nu)^{2m}}\cos 2\pi\nu(s-t). \tag{2.1.5}$$

The eigenvalues of the reproducing kernel are all of multiplicity 2 and are $\lambda_\nu = (2\pi\nu)^{-2m}$, and the eigenfunctions are $\sqrt{2}\,\sin 2\pi\nu t$ and $\sqrt{2}\cos 2\pi\nu t$.

Elements in W_m^0 (per) satisfy the boundary conditions

$$\begin{aligned}\int_0^1 f(u)\,du &= 0,\\ \int_0^1 f^{(k)}(u)\,du &= f^{(k-1)}(1) - f^{(k-1)}(0) = 0, \qquad (2.1.6)\\ & \quad k = 1, \ldots, m.\end{aligned}$$

To remove the condition $\int_0^1 f(u)\,du = 0$, we may adjoin the one-dimensional subspace $\mathcal{H}_0$ spanned by $\{1\}$, and let

$$W_m(\text{per}) = \{1\} \oplus W_m^0(\text{per}).$$

W_m (per), endowed with the norm

$$\|f\|^2 = \left[\int_0^1 f(u)\,du\right]^2 + \int_0^1 (f^{(m)}(u))^2\,du,$$

has the r.k.

$$R(s,t) = 1 + R^1(s,t) = 1 + 2\sum_{\nu=1}^{\infty} \frac{1}{(2\pi\nu)^{2m}} \cos 2\pi\nu(s-t). \qquad (2.1.7)$$

W_m (per) is the subspace of W_m satisfying the periodic boundary conditions $f^{(\nu)}(1) = f^{(\nu)}(0)$, $\nu = 0, 1, 2, \ldots, m-1$.

A closed form expression for $R^1(s,t)$ of (2.1.5) using Bernoulli polynomials was given by Craven and Wahba (1979). Recall that the Bernoulli polynomials $B_r(t)$, $r = 0, 1, \ldots$, $t \in [0,1]$ satisfy the recursion relations

$$B_0(t) = 1$$

$$\frac{1}{r}\frac{d}{dt} B_r(t) = B_{r-1}(t), \quad \int_0^1 B_r(u)\,du = 0, \; r = 1, 2, \ldots.$$

Abramowitz and Stegun (1965) give the formula

$$B_{2m}(x) = (-1)^{m-1} 2(2m)! \sum_{\nu=1}^{\infty} \frac{\cos 2\pi\nu x}{(2\pi\nu)^{2m}}, \qquad x \in [0,1]$$

so that R^1 of (2.1.5) is given by

$$R^1(s,t) = \frac{(-1)^{m-1}}{(2m)!} B_{2m}([s-t])$$

where $[s-t]$ is the fractional part of $s - t$. $R^1(s,t)$ is a stationary covariance on the circle, whose associated stochastic process $X(t)$, $t \in [0,1]$ possess exactly $m-1$ quadratic-mean derivatives and satisfies $D^m X = dW$, and the periodic boundary conditions $X^{(\nu)}(0) = X^{(\nu)}(1)$, $\nu = 0, 1, 2, \ldots, m-1$.

It is instructive to look at the "frequency response" of the smoothing spline in this case. Let n be even and consider

$$y_i = f\left(\frac{i}{n}\right) + \epsilon_i, \quad i = 1, 2, \ldots, n$$

with $f \in W_m$ (per) and ϵ as before. To simplify the argument, we will look at an approximation to the original minimization problem, namely, find f_λ of the form

$$f_\lambda(t) = a_0 + \sum_{\nu=1}^{n/2-1} a_\nu \sqrt{2} \cos 2\pi\nu t + \sum_{\nu=1}^{n/2-1} b_\nu \sqrt{2} \sin 2\pi\nu t + a_{n/2} \cos \pi n t \quad (2.1.8)$$

to minimize

$$\frac{1}{n} \sum_{i=1}^{n} \left(y_i - f\left(\frac{i}{n}\right)\right)^2 + \lambda \int_0^1 (f^{(m)}(u))^2 \, du. \quad (2.1.9)$$

Using the orthogonality relations

$$\begin{aligned}
\frac{2}{n} \sum_{i=1}^{n} \cos 2\pi\nu \frac{i}{n} \cos 2\pi\mu \frac{i}{n} &= 1 \qquad \mu = \nu = 1, \ldots, n/2 - 1, \\
&= 0 \qquad \mu \neq \nu, \ \mu, \nu = 0, 1, \ldots, n/2, \\
\frac{2}{n} \sum_{i=1}^{n} \sin 2\pi\nu \frac{i}{n} \sin 2\pi\mu \frac{i}{n} &= 1 \qquad \mu = \nu = 1, \ldots, n/2 - 1, \\
\frac{1}{n} \sum_{i=1}^{n} \left(\cos 2\pi\nu \frac{i}{n}\right)^2 &= 1, \qquad \nu = 0, n/2, \\
\frac{1}{n} \sum_{i=1}^{n} \cos 2\pi\nu \frac{i}{n} \sin 2\pi\mu \frac{i}{n} &= 0 \qquad \mu \neq \nu,
\end{aligned}$$

we have

$$\begin{aligned}
a_\nu &= \sqrt{\frac{2}{n}} \sum_{i=1}^{n} \cos 2\pi\nu \frac{i}{n} f\left(\frac{i}{n}\right), \qquad \nu = 1, 2, \ldots, n/2 - 1, \\
b_\nu &= \sqrt{\frac{2}{n}} \sum_{i=1}^{n} \sin 2\pi\nu \frac{i}{n} f\left(\frac{i}{n}\right), \qquad \nu = 1, 2, \ldots, n/2 - 1, \\
a_0 &= \sqrt{\frac{1}{n}} \sum_{i=1}^{n} f\left(\frac{i}{n}\right), \\
a_{n/2} &= \sqrt{\frac{1}{n}} \sum_{i=1}^{n} \cos \pi i f\left(\frac{i}{n}\right),
\end{aligned}$$

and letting

$$\hat{a}_\nu = \sqrt{\frac{2}{n}} \sum_{i=1}^{n} y_i \cos 2\pi\nu \frac{i}{n},$$

$$\hat{b}_\nu = \sqrt{\frac{2}{n}}\sum_{i=1}^{n} y_i \sin 2\pi\nu\frac{i}{n},$$
$$\hat{a}_0 = \sqrt{\frac{1}{n}}\sum_{i=1}^{n} y_i,$$
$$\hat{a}_{n/2} = \sqrt{\frac{1}{n}}\sum_{i=1}^{n} y_i \cos \pi i,$$

(2.1.9) becomes

$$\sum_{\nu=0}^{n/2}(a_\nu - \hat{a}_\nu)^2 + \sum_{\nu=1}^{n/2-1}(b_\nu - \hat{b}_\nu)^2 + \lambda\left[\sum_{\nu=1}^{n/2-1}(a_\nu^2 + b_\nu^2)(2\pi\nu)^{2m} + \frac{1}{2}a_{n/2}^2(\pi n)^{2m}\right].$$

The minimizing values are

$$\begin{aligned} a_\nu &= \hat{a}_\nu/(1+\lambda(2\pi\nu)^{2m}), \\ & \qquad\qquad\qquad \nu = 1, 2, \ldots, n/2-1, \\ b_\nu &= \hat{b}_\nu/(1+\lambda(2\pi\nu)^{2m}), \\ a_0 &= \hat{a}_0, \\ a_{n/2} &= \hat{a}_{n/2}/(1+\tfrac{1}{2}\lambda(2\pi\nu)^{2m}), \end{aligned}$$

and

$$f_\lambda(t) = \hat{a}_0 + \sum_{\nu=1}^{n/2-1}\frac{\hat{a}_\nu}{(1+\lambda(2\pi\nu)^{2m})}\cos 2\pi\nu t$$

$$+ \sum_{\nu=1}^{n/2-1}\frac{\hat{b}_\nu}{(1+\lambda(2\pi\nu)^{2m})}\sin 2\pi\nu t$$
$$+ \frac{\hat{a}_{n/2}}{(1+\frac{1}{2}\lambda(\pi n)^{2m})}\cos \pi n t.$$

Thus, the smoothing spline obtained with the penalty functional $\int_0^1 (f^{(m)}(u))^2\,du$ may be viewed as a generalization of the so-called Butterworth filter, which smooths the data by downweighting the component at frequency ν by the weight $\omega(\nu) = (1+\lambda(2\pi\nu)^{2m})^{-1}$.

2.2 Splines on the sphere, the role of the iterated Laplacian.

We will see that the iterated Laplacian plays a role in splines on the circle, the sphere, the line, and the plane and other index sets on which the Laplacian operator commutes with the group operation. In d dimensions the Laplacian is

$$\Delta f = \left(\frac{\partial^2}{\partial x_1^2} + \frac{\partial^2}{\partial x_2^2} + \ldots + \frac{\partial^2}{\partial x_d^2}\right) f \qquad (2.2.1)$$

and the (surface) Laplacian on the (unit) sphere is

$$\Delta f = \frac{1}{\cos^2 \phi} f_{\theta\theta} + \frac{1}{\cos \phi} (\cos \phi f_\phi)_\phi \tag{2.2.2}$$

where θ is the longitude $(0 \le \theta \le 2\pi)$ and ϕ is the latitude $(-\pi/2 \le \phi \le \pi/2)$. Here we use subscripts θ and ϕ to indicate derivatives with respect to θ and ϕ, not to be confused with a subscript λ that indicates dependence on the smoothing parameter λ. On the circle we have

$$\Delta f = \frac{\partial^2}{\partial x^2} f$$

and if $f \in W_m$ (per) then we can integrate by parts to obtain

$$\begin{aligned} \int_0^1 (f^{(m)}(u))^2 \, du &= (-1)^m \int_0^1 f(u) f^{(2m)}(u) \, du \\ &= (-1)^m \int_0^1 f(u) \Delta^m f(u) \, du. \end{aligned} \tag{2.2.3}$$

The eigenfunctions $\{\sqrt{2} \cos 2\pi\nu t, \ \sqrt{2} \sin 2\pi\nu t\}$ of the r.k. R^1 of W_m (per) are the eigenfunctions of the mth iterated Laplacian Δ^m on the circle, while the eigenvalues $\{\lambda_\nu = (2\pi\nu)^{-2m}\}$ are the inverses of the eigenvalues of Δ^m:

$$\Delta^m \Phi_\nu = (-1)^m (2\pi\nu)^{2m} \Phi_\nu,$$

that is,

$$\begin{aligned} D^{2m} \cos 2\pi\nu t &= (-1)^m (2\pi\nu)^{2m} \cos 2\pi\nu t, \\ D^{2m} \sin 2\pi\nu t &= (-1)^m (2\pi\nu)^{2m} \sin 2\pi\nu t. \end{aligned}$$

The generalization to the sphere is fairly immediate. The eigenfunctions of the (surface) Laplacian on the sphere are the spherical harmonics $Y_{\ell s}$, $s = -\ell, \ldots, \ell$, $\ell = 0, 1, \ldots$, where

$$\begin{aligned} Y_{\ell s}(\theta, \phi) &= \theta_{\ell s} \cos s\theta P_\ell^s(\sin \phi), \ \ 0 < s \le \ell, \ \ \ell = 0, 1, \ldots \\ &= \theta_{\ell s} \sin s\theta P_\ell^{|s|}(\sin \phi), \ \ -\ell \le s < 0 \\ &= \theta_{\ell 0} P_\ell(\sin \phi), \ \ s = 0 \end{aligned}$$

where

$$\begin{aligned} \theta_{\ell s} &= \sqrt{2} \sqrt{\frac{2\ell + 1}{4\pi} \frac{(\ell - |s|)!}{(\ell + |s|)!}}, \quad s \ne 0 \\ &= \sqrt{\frac{2\ell + 1}{4\pi}}, \quad s = 0. \end{aligned}$$

$P_\ell, \ell = 0, 1, \ldots$ are the Legendre polynomials and P_ℓ^s are the Legendre functions,

$$P_\ell^s(z) = (1 - z^2)^{s/2} \left(\frac{\partial^s}{\partial z^s} \right) P_\ell(z).$$

The spherical harmonics are the eigenfunctions of the mth iterated Laplacian,

$$\Delta^m Y_{\ell s} = (-1)^m [\ell(\ell+1)]^m Y_{\ell s}, \ s = -\ell, \ldots, \ell, \ \ell = 0, 1, \ldots \tag{2.2.4}$$

and provide a complete orthonormal sequence for $\mathcal{L}_2(\mathcal{S})$, where $\mathcal{S}$ is the unit sphere (see Sansone (1959)).

Let $P = (\theta, \phi)$ and let

$$\begin{aligned} R(P, P') &= 1 + \sum_{\ell=1}^{\infty} \sum_{s=-\ell}^{\ell} \left[\frac{1}{\ell(\ell+1)} \right]^m Y_{\ell s}(P) Y_{\ell s}(P') \\ &= 1 + R^1(P, P'), \ \text{say.} \end{aligned} \tag{2.2.5}$$

Letting

$$f_{\ell s} = \int_S f(P) Y_{\ell s}(P) \, dP,$$

the Hilbert space $\mathcal{H}$ with r.k. R of (2.2.5) is the collection of all functions on the sphere with $f_{00} < \infty$ and

$$\begin{aligned} \|P_1 f\|^2 &= \sum_{\ell=1}^{\infty} \sum_{s=-\ell}^{\ell} f_{\ell s}^2 [\ell(\ell+1)]^m \\ &\equiv \int_S (\Delta^{m/2} f)^2 \, dP < \infty. \end{aligned} \tag{2.2.6}$$

Splines and generalized splines on the sphere have been studied by Wahba (1981d, 1982a), Freeden (1981), and Shure, Parker, and Backus (1982).

There is an addition formula for spherical harmonics analogous to the addition formula for sines and cosines

$$\cos 2\pi\nu s \cos 2\pi\nu t + \sin 2\pi\nu s \sin 2\pi\nu t = \cos 2\pi\nu(t - s),$$

it is

$$\sum_{s=-\ell}^{\ell} Y_{\ell s}(P) Y_{\ell s}(P') = \frac{2\ell+1}{4\pi} P_\ell(\cos \gamma(P, P')) \tag{2.2.7}$$

where γ is the angle between P and P' (see Sansone (1959)). Thus $R^1(P, P')$ collapses to

$$R^1(P, P') = \sum_{\ell=1}^{\infty} \frac{2\ell+1}{4\pi} \frac{1}{[\ell(\ell+1)]^m} P_\ell(\cos \gamma(P, P')), \tag{2.2.8}$$

with the stationarity (dependence only on $\gamma(P, P')$) being evident. Closed form expressions for R^1 for $m = 2$ and (in terms of the dilogarithms) for $m = 3$ were given by Wendelberger (1982), but it appears that closed form expressions for

$$\sum_{\ell=1}^{\infty} \frac{2\ell+1}{[\ell(\ell+1)]^m} P_\ell(z) \tag{2.2.9}$$

are not available for larger m. Reproducing kernels Q^1 that approximate R^1 for $m = 2, 3, \ldots$, and for which closed form expressions are available have been found in Wahba (1981d, 1982a). The eigenvalues $\lambda_{\ell s} = (\ell(\ell+1))^{-m}$ that appear in (2.2.5) and (2.2.8) are replaced by $\xi_{\ell s} = [(\ell+\frac{1}{2})(\ell+1)(\ell+2)\ldots(\ell+2m-1)]^{-1}$, to get

$$\begin{aligned} Q^1(P,P') &= \sum_{\ell=1}^{\infty}\sum_{s=-\ell}^{\ell} \xi_{\ell s} Y_{\ell s}(P) Y_{\ell s}(P') \\ &= \frac{1}{2\pi}\sum_{\ell=1}^{\infty} \frac{1}{(\ell+1)(\ell+2)\ldots(\ell+2m-1)} P_\ell(\cos\gamma(P,P')). \end{aligned}$$

Since

$$\begin{aligned} \xi_{\ell s} &\le \lambda_{\ell s} \le m^{2m}\xi_{\ell s}, \\ Q^1 &\preceq R^1 \preceq m^{2m} Q^1 \end{aligned}$$

where $A \preceq B$ means $B - A$ is nonnegative definite, and $\mathcal{H}_{Q^1}$ and $\mathcal{H}_{R^1}$ are topologically equivalent. Closed form expressions for Q^1 for $m = 2, 5/2, 3, \ldots$, were obtained via the symbol manipulation program MACSYMA.

Another way of computing an approximate spline on the sphere, given noisy data from the model

$$Y_i = f(P_i) + \epsilon_i,$$

is to let f be of the form

$$f = f_{00} + \sum_{\ell=1}^{N}\sum_{s=-\ell}^{\ell} f_{\ell s} Y_{\ell s}, \tag{2.2.10}$$

and choose the $f_{\ell s}$ to minimize

$$\frac{1}{n}\sum_{i=1}^{n}\left(y_i - \sum_{\ell=0}^{N}\sum_{s=-\ell}^{\ell} f_{\ell s} Y_{\ell s}(P_i)\right)^2 + \lambda \sum_{\ell=1}^{N}\sum_{s=-\ell}^{\ell} [(\ell)(\ell+1)]^m f_{\ell s}^2. \tag{2.2.11}$$

Arranging the index set $\{(\ell, s)\}$ in a convenient order, and letting f be the vector of $f_{\ell s}$ and X be the matrix with $i, \ell s$th entry $Y_{\ell s}(P_i)$, we have that (2.2.11) becomes

$$\frac{1}{n}\|y - Xf\|^2 + \lambda f' D f \tag{2.2.12}$$

where D is the diagonal matrix with $\ell s, \ell s$th entry $[(\ell)(\ell+1)]^{-m}$. The minimizing vector f_λ is

$$f_\lambda = (X'X + \lambda D)^{-1} X' y.$$

Methods of computing f_λ for large n and N will be discussed later. Splines on the sphere have found application to the interpolation and smoothing of geophysical and meteorological data. Historical data can be used to choose the $\lambda_{\ell s}$ (see, for example, Stanford (1979)).

2.3 Vector splines on the sphere.

Vector splines on the sphere, for use in smoothing vector fields on the sphere (such as horizontal wind, or magnetic fields), can also be defined. Let the vector field be $\mathbf{V} = (U, V)$ where $U = U(P)$ is the eastward component and $V = V(P)$ is the northward component at P. By the Helmholtz theorem, there exist two functions Ψ and Φ defined on $\mathcal{S}$, called the stream function and the velocity potential, respectively, with the property that

$$U = \frac{1}{a}\left(-\frac{\partial\Psi}{\partial\phi} + \frac{1}{\cos\phi}\frac{\partial\Phi}{\partial\theta}\right), \tag{2.3.1}$$

$$V = \frac{1}{a}\left(\frac{1}{\cos\phi}\frac{\partial\Psi}{\partial\theta} + \frac{\partial\Phi}{\partial\phi}\right),$$

where a is the radius of the sphere. Furthermore, letting the vorticity ζ and the divergence D of $\mathbf{V}$ be defined (as usual) by

$$\zeta = \frac{1}{a\cos\phi}\left[-\frac{\partial}{\partial\phi}(U\cos\phi) + \frac{\partial V}{\partial\theta}\right], \tag{2.3.2}$$

$$D = \frac{1}{a\cos\phi}\left[\frac{\partial U}{\partial\theta} + \frac{\partial}{\partial\phi}(V\cos\phi)\right],$$

we have

$$\zeta = \Delta\Psi, \qquad D = \Delta\Phi, \tag{2.3.3}$$

where now the (surface) Laplacian on the sphere of radius a is

$$\Delta f = \frac{1}{a^2}\left[\frac{1}{\cos^2\phi}f_{\theta\theta} + \frac{1}{\cos\phi}(\cos\phi f_\phi)_\phi\right]. \tag{2.3.4}$$

Ψ and Φ are uniquely determined up to a constant, which we will take to be determined by

$$\int_S \Psi(P)\,dP = \int_S \Phi(P)\,dP = 0. \tag{2.3.5}$$

Given data (U_i, V_i) from the model

$$\begin{aligned} U_i &= U(P_i) + \epsilon_i^U, \quad i = 1, 2, \ldots, n, \\ V_i &= V(P_i) + \epsilon_i^V \end{aligned} \tag{2.3.6}$$

where the ϵ_i^U and ϵ_i^V are random errors, one can define a vector smoothing spline for this data as $\mathbf{V}_{\lambda\delta} = (U_{\lambda,\delta}, V_{\lambda,\delta})$ where

$$\begin{aligned} U_{\lambda,\delta} &= \frac{1}{a}\left(-\frac{\partial\Psi_{\lambda,\delta}}{\partial\phi} + \frac{1}{\cos\phi}\frac{\partial\Phi_{\lambda,\delta}}{\partial\theta}\right), \\ V_{\lambda,\delta} &= \frac{1}{a}\left(\frac{1}{\cos\phi}\frac{\partial\Psi_{\lambda,\delta}}{\partial\theta} + \frac{\partial\Phi_{\lambda\delta}}{\partial\phi}\right), \end{aligned} \tag{2.3.7}$$

and $\Psi_{\lambda,\delta}, \Phi_{\lambda,\delta}$ are the minimizers of

$$\frac{1}{n}\sum_{i=1}^{n}\left(U_i - \frac{1}{a}\left[-\frac{\partial\Psi}{\partial\phi}(P_i) + \frac{1}{\cos\phi}\frac{\partial\Phi}{\partial\theta}(P_i)\right]\right)^2 \tag{2.3.8}$$

$$+\frac{1}{n}\sum_{i=1}^{n}\left(V_i - \frac{1}{a}\left[\frac{1}{\cos\phi}\frac{\partial\Psi}{\partial\theta}(P_i) + \frac{\partial\Phi}{\partial\phi}(P_i)\right]\right)^2$$

$$+\lambda\left[\int_S(\Delta^{m/2}\Psi)^2 dP + \frac{1}{\delta}\int_S(\Delta^{m/2}\Phi)^2 dP\right].$$

$\Delta^{m/2}\Psi$ can, of course, be defined for noninteger $m/2$. If

$$\Psi \sim \sum_{\ell s}\Psi_{\ell s}Y_{\ell s}$$

then

$$\Delta^{m/2}\Psi \sim \sum_{\ell s}[\ell(\ell+1)]^{m/2}\Psi_{\ell s}Y_{\ell s}$$

whenever the sum converges in quadratic mean.

An approximation to the minimizer of (2.3.8) may be obtained by letting Ψ and Φ be of the form

$$\Psi = \sum_{\ell=1}^{N}\sum_{s=-\ell}^{\ell}\alpha_{\ell s}Y_{\ell s}, \tag{2.3.9}$$

$$\Phi = \sum_{\ell=1}^{N}\sum_{s=-\ell}^{\ell}\beta_{\ell s}Y_{\ell s}.$$

Now let X_1 be the matrix with $i,\ell s$th entry $(\partial/\partial\phi)Y_{\ell s}(P)|_{P=P_i}$ and X_2 be the matrix with $i,\ell s$th entry $(1/\cos\phi)(\partial/\partial\theta)Y_{\ell s}(P)|_{P=P_i}$, and let $U = (U_1,\ldots,U_n)$; $V = (V_1,\ldots,V_n)$, then (2.3.8) becomes

$$\frac{1}{n}\|U - \frac{1}{a}(-X_1\alpha + X_2\beta)\|^2$$

$$+\frac{1}{n}\|V - \frac{1}{a}(X_2\alpha + X_1\beta)\|^2$$

$$+\lambda\left[\alpha' D\alpha + \frac{1}{\delta}\beta' D\beta\right]$$

where D is as in (2.2.12).

Given the noisy data U and V, one may estimate the vorticity and divergence as

$$\Delta\Psi_{\lambda,\delta} = \sum_{\ell=1}^{N}\sum_{s=-\ell}^{\ell}\ell(\ell+1)\alpha_{\ell s}^{\lambda,\delta}Y_{\ell s},$$

$$\Delta\Phi_{\ell,s} = \sum_{\ell=1}^{N}\sum_{s=-\ell}^{\ell}\ell(\ell+1)\beta_{\ell s}^{\lambda,\delta}Y_{\ell s}$$

where $\alpha^{\lambda,\delta}$ and $\beta^{\lambda,\delta}$ are the minimizing values of α and β in (2.3.10). This method was proposed in Wahba (1982b); see also Swarztrauber (1981).

2.4 The thin-plate spline on E^d.

The theoretical foundations for the thin-plate spline were laid by Duchon (1975, 1976, 1977) and Meinguet (1979), and some further results and applications to meteorological problems were given in Wahba and Wendelberger (1980). Other applications can be found in Hutchinson and Bischof (1983) and Seaman and Hutchinson (1985). In two dimensions ($d = 2$, $m = 2$, $f = f(x_1, x_2)$), the thin-plate penalty functional is

$$J_2(f) = \int_{-\infty}^{\infty} \int_{-\infty}^{\infty} (f_{x_1x_1}^2 + 2f_{x_1x_2}^2 + f_{x_2x_2}^2)\, dx_1\, dx_2 \tag{2.4.1}$$

and, in general,

$$J_m(f) = \sum_{\nu=0}^{m} \int_{-\infty}^{\infty} \int_{-\infty}^{\infty} \binom{m}{\nu} \left(\frac{\partial^m f}{\partial x_1^{\nu} \partial x_2^{m-\nu}} \right)^2 dx_1\, dx_2. \tag{2.4.2}$$

For $d = 3$, $m = 2$, the thin-plate penalty functional is

$$J_2(f) = \int_{-\infty}^{\infty} \int_{-\infty}^{\infty} (f_{x_1x_1}^2 + f_{x_2x_2}^2 + f_{x_3x_3}^2 + 2[f_{x_1x_2}^2 + f_{x_1x_3}^2 + f_{x_2x_3}^2])\, dx_1\, dx_2\, dx_3 \tag{2.4.3}$$

and the form for general d, m is

$$J_m^d(f) = \sum_{\alpha_1 + \ldots + \alpha_d = m} \frac{m!}{\alpha_1! \ldots \alpha_d!} \int_{-\infty}^{\infty} \cdots \int_{-\infty}^{\infty} \cdot \left(\frac{\partial^m f}{\partial x_1^{\alpha_1} \ldots \partial x_d^{\alpha_d}} \right)^2 \prod_j dx_j. \tag{2.4.4}$$

A formula analogous to (2.2.3) holds here.

Letting

$$< f, g >= \sum_{\alpha_1 + \ldots + \alpha_d = m} \frac{m!}{\alpha_1! \ldots \alpha_d!} \int_{-\infty}^{\infty} \cdots \int_{-\infty}^{\infty} \cdot \left(\frac{\partial^m f}{\partial x_1^{\alpha_1} \ldots dx_d^{\alpha_d}} \right) \left(\frac{\partial^m g}{\partial x_1^{\alpha_1} \ldots \partial x_d^{\alpha_d}} \right) \prod_j dx_j, \tag{2.4.5}$$

we note that a formal integration by parts results in

$$< f, g >= \int_{-\infty}^{\infty} \cdots \int_{-\infty}^{\infty} f \cdot \Delta^m g + \mathcal{B}, \tag{2.4.6}$$

where $\mathcal{B}$ represents boundary values at infinity.

We will suppose $f \in \mathcal{X}$, a space of functions whose partial derivatives of total order m are in $\mathcal{L}_2(E^d)$ (see Meinguet (1979) for more details on $\mathcal{X}$). We want $\mathcal{X}$

endowed with the seminorm $J_m^d(f)$ to be an r.k.h.s., that is, we want to have the evaluation functionals be bounded in $\mathcal{X}$. For this it is necessary and sufficient that $2m - d > 0$.

Now, let the data model be

$$y_i = f(x_1(i), \ldots, x_d(i)) + \epsilon_i \;, \; i = 1, \ldots, n, \tag{2.4.7}$$

where $f \in \mathcal{X}$ and $\epsilon = (\epsilon_1, \ldots, \epsilon_n)' \sim \mathcal{N}(0, \sigma^2 I)$. A thin-plate smoothing spline is the solution to the following variational problem. Find $f \in \mathcal{X}$ to minimize

$$\frac{1}{n}\sum_{i=1}^{n}(y_i - f(x_1(i), \ldots, x_d(i)))^2 + \; \lambda J_m^d(f). \tag{2.4.8}$$

We will use the notation $t = (x_1, \ldots, x_d)$ and $t_i = (x_1(i), \ldots, x_d(i))$. The null space of the penalty functional J_m^d is the $M = \begin{pmatrix} d+m-1 \\ d \end{pmatrix}$-dimensional space spanned by the polynomials in d variables of total degree $\le m-1$. For example, for $d = 2, \; m = 2$, then $M = 3$ and the null space is spanned by ϕ_1, ϕ_2, and ϕ_3 given by

$$\phi_1(t) = 1, \;\; \phi_2(t) = x_1, \;\; \phi_3(t) = x_2.$$

In general, we will denote the M monomials of total degree less than m by $\phi_1, \ldots, \phi_M$.

Duchon (1977) showed that, if $t_1, \ldots, t_n$ are such that least squares regression on $\phi_1, \ldots, \phi_M$ is unique, then (2.4.8) has a unique minimizer f_λ, with representation

$$f_\lambda(t) = \sum_{\nu=1}^{M} d_\nu \phi_\nu(t) + \sum_{i=1}^{n} c_i E_m(t, t_i), \tag{2.4.9}$$

where E_m is a Green's function for the m-iterated Laplacian. Letting

$$\begin{aligned} E(\tau) &= \theta_{m,d}|\tau|^{2m-d}\ln|\tau| && \text{if } 2m-d \text{ an even integer,} \\ &= \theta_{m,d}|\tau|^{2m-d} && \text{otherwise,} \end{aligned} \tag{2.4.10}$$

where

$$\theta_{m,d} = \frac{(-1)^{d/2+1+m}}{2^{2m-1}\pi^{d/2}(m-1)!(m-d/2)!} \qquad \text{if } 2m-d \text{ is an even integer,}$$

$$\theta_{m,d} = \frac{\Gamma(d/2-m)}{2^{2m}\pi^{d/2}(m-1)!} \qquad \text{otherwise,}$$

and letting $|t - t_i| = (\sum_{j=1}^{d}(x_j - x_j(i))^2)^{1/2}$, $E_m(s,t)$ is given by

$$E_m(s,t) = E(|s-t|). \tag{2.4.11}$$

Formally,

$$\Delta^m E_m(\cdot, t_i) = \delta_{t_i}, \tag{2.4.12}$$

where δ_{t_i} is the Dirac delta function, so that

$$\Delta^m f_\lambda(t) = 0 \text{ for } t \neq t_i,\ i = 1, \ldots, n, \tag{2.4.13}$$

analogous to the univariate polynomial spline case where $(\partial^{2m}/\partial x^{2m}) f_\lambda(x) = 0$ for $x \neq x_1, \ldots, x_n$. The functions $E_m(t, t_i)$, $i = 1, \ldots, n$, play the same role as $\xi_i(t) = R^1(t_i, t)$ in Section 1.3, except that $E_m(\cdot, \cdot)$ is not positive definite. $E_m(\cdot, \cdot)$ is *conditionally positive definite*, a property that turns out to be enough. To explain the notion of conditional positive definiteness, we need the notion of a generalized divided difference. Given $t_1, \ldots, t_n \in E^d$, let T be the $n \times M$ matrix with $i\nu$th entry $\phi_\nu(t_i)$. In one dimension, T is always of full column rank if the t_i's are distinct. In two and higher dimensions it must be an explicit assumption that T is of full column rank, which we will always make. If, for example, $t_1, \ldots, t_n$ fall on a straight line on the plane this assumption will fail to hold. Now let $c \in E^n$ be any vector satisfying $T'c = 0$. Then $(c_1, \ldots, c_n)$, associated with $t_1, \ldots, t_n$, is called a generalized divided difference (g.d.d.) of order m, since it annihilates all polynomials of total degree less than m, that is, $\sum_{i=1}^n c_i \phi_\nu(t_i) = 0$, $\nu = 1, 2, \ldots, M$. Recall that the ordinary first-order divided differences are of the form $(f(t_{i+1}) - f(t_i))/(t_{i+1} - t_i)$ and annihilate constants, second-order divided differences are of the form

$$\left(\frac{f(t_{i+2}) - f(t_{i+1})}{t_{i+2} - t_{i+1}} - \frac{f(t_{i+1}) - f(t_i)}{t_{i+1} - t_i} \right) / (t_{i+2} - t_i)$$

and annihilate constants and linear functions, and so forth, thus a g.d.d. is a generalization of an ordinary divided difference.

Duchon (1977) and Matheron (1973) both have proved the following: Given $t_1, \ldots, t_n$ such that T is of rank M, let $K_{n \times n}$ be the $n \times n$ matrix with ijth entry $E_m(t_i, t_j)$. Then

$$c'Kc > 0 \tag{2.4.14}$$

for any g.d.d. c of order m, that is, for any c such that $T'c = 0$. E_m is then called m-conditionally (strictly) positive definite.

Now, let $E_t(\cdot) = E_m(t, \cdot)$ and write

$$\begin{aligned} < E_t, E_s > \quad = \quad & \sum_{\alpha_1 + \ldots + \alpha_d = m} \frac{m!}{\alpha_1! \ldots \alpha_d!} \\ & \cdot \int_{-\infty}^{\infty} \cdots \int_{-\infty}^{\infty} \frac{\partial^m E_t}{\partial x_1^{\alpha_1} \ldots \partial x_d^{\alpha_d}} \cdot \frac{\partial^m E_s}{\partial x_1^{\alpha_1} \ldots \partial x_d^{\alpha_d}} \prod_j dx_j. \end{aligned} \tag{2.4.15}$$

A formal integration by parts yields

$$\begin{aligned} \int_{-\infty}^{\infty} \cdots \int_{-\infty}^{\infty} E_t(u) \Delta^m E_s(u)\, du + \mathcal{B} \quad &= \quad E_t(s) + \mathcal{B} \\ &= \quad E_m(s, t) + \mathcal{B}. \end{aligned} \tag{2.4.16}$$

This calculation is not legitimate, since the boundary values at ∞ will be infinite. However, it is known from the work of Duchon and Meinguet that if we let

$$g(s) = \sum_{i=1}^{n} c_i E_m(s, t_i)$$

where $c = (c_1, \ldots, c_n)'$ is a g.d.d., then g has appropriate behavior at infinity and we can write

$$\begin{aligned} < g, g > &= < \sum_{i=1}^{n} c_i E_m(\cdot, t_i), \sum_{j=1}^{n} c_j E_m(\cdot, t_j) > \\ &= \sum_{i,j} c_i c_j \, E_m(t_i, t_j) \\ &= c'Kc > 0. \end{aligned} \tag{2.4.17}$$

By substituting (2.4.9) into (2.4.8) and using (2.4.17), we obtain that c, d are the minimizers of

$$\frac{1}{n}\|y - Td - Kc\|^2 + \lambda c'Kc \tag{2.4.18}$$

subject to $T'c = 0$. To find the minimizers c and d of this expression, we let the QR decomposition of T be

$$T = (Q_1 : Q_2)\begin{pmatrix} R \\ 0 \end{pmatrix} \tag{2.4.19}$$

where $(Q_1 : Q_2)$ is orthogonal and R is lower triangular. Q_1 is $n \times M$ and Q_2 is $n \times (n - M)$. Since $T'c = 0$, c must be in the column space of Q_2, $c = Q_2\gamma$ for some $m - M$ vector γ. By the orthogonality of $(Q_1 : Q_2)$ we have $\|x\|^2 = \|Q_1'x\|^2 + \|Q_2'x\|^2$ for any $x \in E^n$. Using this and substituting $Q_2\gamma$ for c in (2.4.18) gives

$$\frac{1}{n}\|Q_2'y - Q_2'KQ_2\gamma\|^2 + \frac{1}{n}\|Q_1'y - Rd - Q_1'KQ_2\gamma\|^2 + \lambda\gamma'Q_2'KQ_2\gamma. \tag{2.4.20}$$

It is seen that the minimizers d and γ satisfy

$$Rd = Q_1'(y - KQ_2\gamma) \tag{2.4.21}$$

and

$$Q_2'y = (Q_2'KQ_2 + \lambda I)\gamma. \tag{2.4.22}$$

These relations can be seen to be equivalent to

$$\begin{aligned} Mc + Td &= y, & (2.4.23) \\ T'c &= 0 & (2.4.24) \end{aligned}$$

with $M = K + n\lambda I$, by multiplying (2.4.23) by Q_2' and letting $c = Q_2'\gamma$. The columns of Q_2 are all the g.d.d.'s.

It is possible to come to the same result for the minimizer of (2.4.8) via reproducing kernels. Let $s_1, \ldots, s_M$ be any M fixed points in E^d such that least squares regression on the M-dimensional space of polynomials of total degree less than m at the points $s_1, \ldots, s_M$ is unique, that is, the $M \times M$ matrix S, with $i\nu$th entry $\phi_\nu(s_i)$ is of full rank. (In this case we call the points $s_1, \ldots, s_M$ unisolvent.) Let $p_1, \ldots, p_M$ be the (unique) polynomials of total degree less than m satisfying $p_i(s_j) = 1,\ i = j, = 0, i \neq j$, and let

$$\begin{aligned} R^1(s,t) = E_m(s,t) \quad &- \quad \sum_{\nu=1}^{M} p_\nu(t) E_m(s_\nu, s) \\ &- \quad \sum_{\mu=1}^{M} p_\mu(s) E_m(t, s_\mu) \\ &+ \sum_{\mu,\nu=1}^{M} p_\mu(s) p_\nu(t) E_m(s_\mu, s_\nu). \end{aligned} \tag{2.4.25}$$

Letting

$$R_t^1(\cdot) = R^1(\cdot, t) \tag{2.4.26}$$

we have

$$R_t^1(s) = E_m(s,t) - \sum_{\nu=1}^{M} p_\nu(t) E_m(s_\nu, s) + \pi_t(s) \tag{2.4.27}$$

where for fixed t, $\pi_t(\cdot)$ is a polynomial of degree $m-1$ in s. Now for fixed t, consider the points $(t, s_1, \ldots, s_M)$ and coefficients $(1, -p_1(t), \ldots, -p_M(t))$. These coefficients together with the points $(t, s_1, \ldots, s_M)$ constitute a g.d.d., since

$$\phi(t) - \sum_{\nu=1}^{M} \phi(s_\nu) p_\nu(t) \equiv 0 \tag{2.4.28}$$

for any polynomial ϕ of total degree less than m. Equation (2.4.28) follows since the sum in (2.4.28) is a polynomial of degree less than m that interpolates to ϕ at a set of M unisolvent points, therefore it must be zero. In fact,

$$< R_t^1, R_s^1 >= R^1(s,t).$$

$R^1(s_\nu, s_\nu) = 0,\ \nu = 1, 2, \ldots, M$, but R^1 is positive semidefinite. R^1 is an r.k. for $\mathcal{H}_1$, the subspace of $\mathcal{X}$ of codimension M of functions satisfying $f(s_\nu) = 0,\ \nu = 1, 2, \ldots, M$, and $\mathcal{X}$ is the direct sum of $\mathcal{H}_0 = \text{span}\ \{p_1, \ldots, p_M\}$ and $\mathcal{H}_1$ with $J_m^d(f) = \|P_1 f\|^2$. It follows from Section 1.3 that f_λ has a representation

$$f_\lambda(t) = \sum_{\nu=1}^{M} d_\nu \phi_\nu(t) + \sum_{i=1}^{n} c_i R_{t_i}^1(t) \tag{2.4.29}$$

for some d, c, with $T'c = 0$. The end result from (2.4.29) can be shown to be the same as (2.4.9) since $\sum_{i=1}^n c_i E_m(t, t_i)$ and $\sum_{i=1}^n c_i R_{t_i}^1(t)$ differ by a polynomial of total degree less than m in t if c is a g.d.d.

2.5 Another look at the Bayes model behind the thin-plate spline.

Returning to (1.5.7), consider the "fixed effects" model

$$Y_i = F(t_i) + \epsilon_i, \ i = 1, 2, \ldots, n$$

where

$$F(t) = \sum_{\nu=1}^{M} \theta_\nu \phi_\nu(t) + b^{1/2} X(t), \ t \in \mathcal{T}. \tag{2.5.1}$$

One such model that will result in the thin-plate spline is: $\{\phi_1, \ldots, \phi_M\}$ span $\mathcal{H}_0$, the space of polynomials of total degree less than m, and

$$EX(s)X(t) = R^1(s,t),$$

with $R^1(s,t)$ given by (2.4.25). The points $s_1, \ldots, s_M$ used in defining R^1 were arbitrary, and it can be seen that it is not, in fact, necessary to know the entire covariance of $X(t), \ t \in \mathcal{T}$. Any covariance for which the g.d.d.'s of X satisfy

$$E \sum_l c_l X(t_l) \sum_k c_k X(t_k) = \sum_{lk} c_l c_k E_m(t_l, t_k)$$

whenever $\sum_\ell c_\ell \phi_\nu(t_l) = 0, \nu = 1, 2, \ldots, M$, will result in the *same* thin-plate spline. Looking at this phenomenon from another point of view, one can replace $X(t), t \in \mathcal{T}$ in the model (2.5.1) by $\tilde{X}(t) = X(t) + \sum_{\nu=1}^{M} \tilde{\theta}_\nu \phi_\nu(t)$ where the $\tilde{\theta}_\nu$ are arbitrary, without changing the model. The estimation procedure assigns as much of the "explanation" of the data vector as possible to θ in (2.5.1) and not $\tilde{\theta}$. This kind of reasoning was behind the development of "kriging" due to a South African mining engineer, David Krige (see the references in Delfiner (1975)). The motivation for Krige's work was to estimate the total ore content of a volume of earth from observations from core samples. It was assumed that the ore density was a random process $Y(t), t \in E^d$, whose generalized divided differences were stationary, and that it had a so-called variogram $\tilde{E}(\tau)$ with the property that

$$E \sum_l c_l Y(s_l) \sum_k c_k Y(s_k) = \sum_l \sum_k c_l c_k \tilde{E}|s_l - s_k|$$

whenever the $\{c_l, s_l\}$ constituted a g.d.d., and a "drift," or mean-value function, of the form

$$\sum_{\nu=1}^{M} \theta_\nu \phi_\nu.$$

The kriging estimate $\hat{Y}(t), \ t \in T$, was defined as the minimum variance, conditionally unbiased (with respect to θ) linear estimate of $Y(t)$ given $Y_i = y_i$, and if $\hat{Y}(t)$ is the estimate of $Y(t)$, then $\int_\Omega \hat{Y}(t)\,dt$ is the conditionally unbiased, minimum variance estimate of $\int_\Omega Y(t)\,dt$ (compare Section 1.5).

This connection between spline estimation and kriging was demonstrated in Kimeldorf and Wahba (1971, §7,) although the word kriging was never mentioned. We had not heard of it at the time. Duchon (1975, 1976) gave a general version of this result in French, and various connections between the two lines of research, which have been carried out fairly independently until the last few years, have been rediscovered a number of times.

Matheron (1973) characterized the class of k-conditionally positive-definite functions on E^d, in particular, letting

$$K(\tau) = \sum_{p=0}^{k} (-1)^{p+1} a_p \tau^{2p+1} \tag{2.5.2}$$

where the coefficients a_p satisfy

$$\sum_{p=0}^{k} \frac{a_p}{\pi^{2p+2+d/2}} \frac{\Gamma(\frac{1}{2}(2p+1+d))}{\Gamma[1+\frac{1}{2}(2p+1)]} \rho^{-d-2p+1} \geq 0 \tag{2.5.3}$$

for any $\rho \geq 0$. Matheron showed that

$$E(s,t) = K(|s-t|)$$

is k-conditionally positive definite. (Note that if $E(s,t)$ is k-conditionally positive definite, it is $k+1$ conditionally positive definite.) Much of the work on kriging involves variograms of the form

$$E(s,t) = |s-t|^3 - \beta|s-t|$$

or

$$E(s,t) = |s-t|^5 - \beta_1|s-t|^3 + \beta_2|s-t|,$$

where the β's are estimated from the data. See Delfiner (1975), Journel and Huijbregts (1978), and Cressie and Horton (1987).

We will now make some remarks concerning the variational problem associated with generalized covariances of the form $E(s,t) = K(|s-t|)$, where K is as in (2.5.2). It is easy to see what happens to the analogous case on the circle. Letting $s,t \in [0,1]$, let

$$\mathcal{E}_m(s,t) = 2\sum_{\nu=1}^{\infty} \frac{\cos 2\pi\nu(s-t)}{(2\pi\nu)^{2m}}$$

and let

$$R(s,t) = \sum_{l=m}^{m+k} \alpha_{l-m} \mathcal{E}_l(s,t) \tag{2.5.4}$$

with $\alpha_0 = 1$. Then the eigenvalues of $\mathcal{E}_m$ are $\lambda_\nu(\mathcal{E}_m) = (2\pi\nu)^{-2m}$ and the eigenvalues $\lambda_\nu(R)$ of R are

$$\lambda_\nu(R) = \sum_{l=m}^{m+k} \frac{\alpha_{l-m}}{(2\pi\nu)^{2l}} = \frac{1}{(2\pi\nu)^{2(m+k)}} \sum_{j=0}^{k} \alpha_{k-j}(2\pi\nu)^{2j}. \tag{2.5.5}$$

In order for R of (2.5.4) to be a covariance, it is necessary that $\lambda_\nu(R) \geq 0$, for this it is sufficient that $\sum_{j=0}^k \alpha_{k-j}\rho^{2j} \geq 0$ for all $\rho > 0$. In the discussion below, we will assume that the α_j's are such that $\lambda_\nu(R) > 0$. Letting $f_\nu^2 = c_\nu^2 + s_\nu^2$ where $c_\nu = \sqrt{2}\int f(t)\cos 2\pi\nu t$, $s_\nu = \sqrt{2}\int f(t)\sin 2\pi\nu t$, then we have that the squared norms associated with $\mathcal{E}_m$ and R are, respectively,

$$\sum_{\nu=1}^{\infty}(2\pi\nu)^{2m} f_\nu^2 \tag{2.5.6}$$

and

$$\sum_{\nu=1}^{\infty}\frac{(2\pi\nu)^{2(m+k)}}{\sum_{j=0}^k \alpha_{k-j}(2\pi\nu)^{2j}} f_\nu^2 = \sum_{\nu=1}^{\infty}(2\pi\nu)^{2m}H(\nu)f_\nu^2 \tag{2.5.7}$$

where

$$H(\nu) = \frac{(2\pi\nu)^{2k}}{\sum_{j=0}^k \alpha_{k-j}(2\pi\nu)^{2j}} = \left(1 + \sum_{j=1}^k \frac{\alpha_j}{(2\pi\nu)^{2j}}\right)^{-1}.$$

As $\nu \to \infty$, $H(\nu) \to 1$. If $\sum_{j=1}^k(\alpha_j/(2\pi\nu)^{2j})$ is bounded strictly above -1, then the two norms satisfy

$$a\|f\|^2_{\mathcal{H}_{\mathcal{E}_m}} \leq \|f\|^2_{\mathcal{H}_R} \leq b\|f\|^2_{\mathcal{H}_{\mathcal{E}_m}}$$

for some $0 < a \leq b < \infty$. Then $f \in \mathcal{H}_R$ if and only if $f \in \mathcal{H}_{\mathcal{E}_m}$ and the two spaces are topologically equivalent.

We now return to E^d and the thin plate penalty functional. Since we are running out of symbols we will use $(\alpha_1, \ldots, \alpha_d)$ as a multi-index below, not to be confused with the α's in the definition of R above. Observing that the Fourier transform of $\partial^m f/\partial x_1^{\alpha_1}\ldots\partial x_d^{\alpha_d}$ is

$$\widehat{\frac{\partial^m f}{\partial x_1^{\alpha_1}\ldots\partial x_d^{\alpha_d}}} = \prod_{l=1}^d (2\pi i w_l)^{\alpha_l}\hat{f}(w_1,\ldots,w_d)$$

where "$\hat{f}$" denotes Fourier transform and that

$$(|w_1|^2 + \ldots + |w_d|^2)^m = \sum_{\alpha_1+\ldots+\alpha_d}\frac{m!}{\alpha_1!\ldots\alpha_d!}\prod_{l=1}^d |w_l|^{2\alpha_l}$$

we have

$$J_m^d(f) = \int_{-\infty}^{\infty}\cdots\int_{-\infty}^{\infty}\|2\pi w\|^{2m}|\hat{f}(w)|^2\prod_l dw_l. \tag{2.5.8}$$

The argument below is loosely adapted from a recent thesis by Thomas-Agnan (1987), who considered general penalty functionals of the form

$$\int \alpha^2(w)|\hat{f}(w)|^2\,dw.$$

The penalty functional associated with the variogram $\sum_{l=m}^{m+k} \alpha_{l-m} E_l$ where E_l is defined as in (2.4.11) must be

$$\tilde{J}_m^d(f) = \int \cdots \int \frac{\|2\pi w\|^{2(m+k)}}{\sum_{j=0}^k \alpha_{k-j} \|2\pi w\|^{2j}} |\hat{f}(w)|^2 \prod_l dw_l. \qquad (2.5.9)$$

This can be conjectured by analogy to (2.5.7). For rigorous details in the $d = 1$ case, see Thomas-Agnan (1987).

In going from the circle to E^d we must be a little more careful, however. Letting

$$H(w) = \frac{\|2\pi w\|^{2k}}{\|2\pi w\|^{2k} + \alpha_1 \|2\pi w\|^{2(k-1)} + \ldots + \alpha_k},$$

we have

$$\tilde{J}_m^d(f) = \int \cdots \int H(w) \|2\pi w\|^{2m} |\hat{f}(w)|^2 dw,$$

since $H(w) \to 1$ as $w \to \infty$, the tail behavior of the Fourier transforms of functions for which J_m^d and $\tilde{J}_m^d$ are finite will be the same. This tail behavior ensures that the total derivatives of order m are in $\mathcal{L}_2$. However, we have $\lim_{w \to 0} H(w) = 0$, in particular, $\lim_{w \to 0} (H(w) \|2\pi w\|^{2m}) / \|2\pi w\|^{2(m+k)} \to 1/\alpha_k$. It can be argued heuristically that the polynomials of total degree less than $m+k$ are in the null space of $\tilde{J}_m^d(f)$, by writing

$$\begin{aligned} \tilde{J}_m^d(f) &= \int \cdots \int \frac{\|2\pi w\|^{2(m+k)} |\hat{f}(w)|^2}{\|2\pi w\|^{2k} + \alpha_1 \|2\pi w\|^{2(k-1)} + \ldots + \alpha_k} \prod_l dw_l \\ &= \int \cdots \int \frac{\|\widehat{\partial^{2(m+k)} f}\|^2}{\|2\pi w\|^{2k} + \alpha_1 \|2\pi w\|^{2(k-1)} + \ldots + \alpha_k} \prod_l dw_l \end{aligned} \qquad (2.5.10)$$

where $\widehat{\partial^{2(m+k)}} f$ is the Fourier transform of the $2(m+k)$th total derivative of f. If (2.5.10) is valid then $\tilde{J}_m^d(f) = 0$ for f a polynomial of total degree less than $m + k$.

Let $\tilde{s}_1, \ldots, \tilde{s}_{\tilde{M}}$ be a unisolvent set of

$$\tilde{M} = \binom{d + (m+k) - 1}{d}$$

points in E^d, and let

$$E(s,t) = \sum_{l=m}^{m+k} \alpha_{l-m} E_l(s,t)$$

where the α's satisfy conditions ensuring that $E(s,t)$ is $m + k$ conditionally positive definite. Let $p_1, \ldots, p_{\tilde{M}}$ be the $\tilde{M}$ polynomials satisfying $p_i(\tilde{s}_j) = 1$, $i =$

j, and 0, $i \neq j$, and let

$$\begin{aligned} R^1(s,t) = E(s,t) &- \sum_{\nu=1}^{\tilde{M}} \tilde{p}_\nu(t) E(\tilde{s}_\nu, s) \\ &- \sum_{\mu=1}^{\tilde{M}} \tilde{p}_\mu(s) E(t, \tilde{s}_\mu) \\ &+ \sum_{\mu,\nu=1}^{\tilde{M}} \tilde{p}_\mu(s) \tilde{p}_\nu(t) E(\tilde{s}_\mu, \tilde{s}_\nu). \end{aligned}$$

One can argue analogously to Section 2.4 that R^1 must be a positive-definite function that has the reproducing kernel property under the norm defined by $\tilde{J}_m^d(f)$.

CHAPTER 3

Equivalence and Perpendicularity, or, What's So Special About Splines?

3.1 Equivalence and perpendicularity of probability measures.

In Section 1.2 we considered a penalty functional that is a seminorm in W_m and a penalty functional that is a seminorm in $\tilde{W}_m$, a topologically equivalent space. Note that it took several additional parameters (the a_i's) to specify the seminorm in $\tilde{W}_m$. Aside from some computational advantages (considerable in the case of polynomial splines) why, in practical work, should we choose one penalty functional over another?

Continuing with this inquiry, one may ask if there is a particular reason for using a kriging estimator with $K(\tau)$ of (2.5.2) given by

$$K(\tau) = \sum_{p=0}^{k} (-1)^{p+1} a_p \tau^{2p+1},$$

where the a_p must be estimated from the data, rather than the thin-plate spline estimate, that corresponds to the simpler

$$K(\tau) = (-1)^{p+1} a_p \tau^{2p+1}.$$

How far should one go in estimating parameters in the penalty functional when valid prior information is not otherwise available? The theory of equivalence and perpendicularity gives an answer to this question. We will now describe the results we need.

A probability measure P_1 is said to dominate another measure P_2 ($P_1 \succ P_2$) if $P_1(A) = 0 \Rightarrow P_2(A) = 0$. P_1 is said to be equivalent to P_2 ($P_1 \equiv P_2$) if $P_2 \prec P_1$ and $P_2 \succ P_1$. P_1 is said to be perpendicular to P_2 if there exists an event A such that $P_1(A) = 0$ and $P_2(A) = 1$. It is known that Gaussian measures are either equivalent or perpendicular. Any two nontrivial Gaussian measures on E^1 are equivalent, and two Gaussian measures on E^d are equivalent if the null spaces of their covariance matrices coincide, otherwise they are perpendicular.

Considering Gaussian measures on infinite sequences of zero-mean independent random variables $X_1, X_2, \ldots$, Hajek (1962a,b) has given necessary and sufficient conditions for equivalence. Let $EX_\nu^2 = \sigma_\nu^2(1)$ under P_1 and $\sigma_\nu^2(2)$ under P_2. If $\sigma_\nu^2(1) > 0$ and $\sigma_\nu^2(2) = 0$ or $\sigma_\nu^2(1) = 0$ and $\sigma_\nu^2(2) > 0$ for some ν, then the two processes are perpendicular. Suppose $\sigma_\nu^2(1)$ and $\sigma_\nu^2(2)$ are positive or zero together. Then $P_1 \equiv P_2$ if and only if $\sum_{\nu=1}^\infty (1 - \sigma_\nu^2(1)/\sigma_\nu^2(2))^2 < \infty$.

Let us now consider a stochastic process $X(t),\ t \in \mathcal{T}$ with the Karhunen–Loeve expansion (see Section 1.1)

$$X(t) = \sum_{\nu=1}^{\infty} X_\nu \, \Psi_\nu(t), \tag{3.1.1}$$

where $\{\Psi_\nu\}$ is an orthonormal sequence in $\mathcal{L}_2(\mathcal{T})$ and the $\{X_\nu\}$ are independent, zero-mean Gaussian random variables with $EX_\nu^2 = \sigma_\nu^2(i) > 0$ under $P_i,\ i = 1, 2$. Then it follows from Hajek's result that P_1 and P_2 will be equivalent or perpendicular accordingly as $\sum_{\nu=1}^\infty (1 - \sigma_\nu^2(1)/\sigma_\nu^2(2))^2$ is finite or infinite.

Now consider the following example of Section 2.1:

$$\sigma_\nu^2(1) = b_1(2\pi\nu)^{-2m_1}, \qquad \sigma_\nu^2(2) = b_2(2\pi\nu)^{-2m_2},$$

P_1 and P_2 will be equivalent if $b_1 = b_2$ and $m_1 = m_2$, and perpendicular otherwise. (Here m_1 and m_2 need not be integers.)

Suppose we have a prior distribution with $\sigma_\nu^2 = b_*(2\pi\nu)^{-2m_*}$, where b_* and m_* are *unknown*. The perpendicularity fact above means that we can expect to find a consistent estimator for (b, m) given $X(t_1), \ldots, X(t_n)$ as $t_1, \ldots, t_n$ become dense in $\mathcal{T}$. To see this, let (b_*, m_*) be any fixed value of (b, m). Then there exists a set $A(b_*, m_*)$ in the sigma field for $\{X(t),\ t \in \mathcal{T}\}$, equivalently in the sigma field for $\{X_1, X_2, \ldots\}$ such that $P(\{X_1, X_2, \ldots\} \in A(b_*, m_*)) = 1$ if (b_*, m_*) is true and zero otherwise. Thus the estimate is formed by determining in which $A(b, m)$ $\{X_1, X_2, \ldots\}$ lies. Under some mild regularity conditions (for example, $X(t),\ t \in \mathcal{T}$ continuous in quadratic mean), it is sufficient to observe $X(t)$ only for t in a dense subset of $\mathcal{T}$.

Now consider the case with $\sigma_\nu^2(1) = (2\pi\nu)^{-2m}$ and $\sigma_\nu^2(2) = [(2\pi\nu)^{2m} + \theta^2(2\pi\nu)^{2(m-1)}]^{-1}$. By considering the problem in its usual complex form (we omit the details), the $\{\sigma_\nu^2(2)\}$ can be shown to be the eigenvalues associated with the penalty $\int_0^1 [f^{(m)}(t) + \theta f^{(m-1)}(t)]^2\, dt$ for the periodic spline case of Section 2.1.

Then

$$\frac{\sigma_\nu^2(1)}{\sigma_\nu^2(2)} = 1 + \frac{\theta^2}{(2\pi\nu)^2}$$

and

$$\sum_{\nu=1}^{\infty} \left(1 - \frac{\sigma_\nu^2(1)}{\sigma_\nu^2(2)}\right)^2 = \sum_{\nu=1}^{\infty} \left[\frac{\theta^2}{(2\pi\nu)^2}\right]^2 < \infty$$

so that P_1 and P_2 are equivalent. This means that there *cannot* be a consistent estimate of θ, since if there were we would know θ "perfectly" (w.pr.1) given $X(t),\ t \in \mathcal{T}$, and then we could tell w.pr.1 which of P_1 or P_2 is true, which contradicts the fact that they are equivalent.

Hajek (1962a) considers the nonperiodic case on a finite interval of the real line. The result, loosely stated, is that if $X(t)$ formally satisfies

$$\sum_{j=0}^{m} a_{m-j}^{(i)} X^{(j)} = dW \tag{3.1.2}$$

under P_i, $i = 1, 2$ and the boundary random variables are equivalent, then $P_1 \equiv P_2$ if $a_0^{(1)} = a_0^{(2)}$ and $P_1 \perp P_2$ if $a_0^{(1)} \neq a_0^{(2)}$. Thus $a_1, \ldots, a_{m-1}$ cannot be estimated consistently from data on a finite interval. More generally, if X is the restriction to a finite interval of a stationary Gaussian process on the real line with spectral density

$$f(w) = \left| \frac{\sum_{k=0}^{q(i)} b_{q-k}^{(i)} (iw)^k}{\sum_{k=0}^{p(i)} a_{p-k}^{(i)} (iw)^k} \right|^2, \qquad i = 1, 2$$

then $P_1 \equiv P_2$ if $q(1) - p(1) = q(2) - p(2)$ and $a_0(1)/b_0(1) = a_0(2)/b_0(2)$ and $P_1 \perp P_2$ otherwise (see Hajek (1962b, Thm. 2.3) for further details). Parzen (1963) discusses conditions for the equivalence and perpendicularity in terms of the properties of reproducing kernels.

3.2 Implications for kriging.

Now we will examine the periodic version of the prior related to kriging, from the d-dimensional version of (2.5.4), to see which coefficients in the variogram should be consistently estimable from data on a bounded region. To see what happens most easily in the d-dimensional periodic case, it is convenient to think of the eigenfunctions of the reproducing kernel in complex form. Letting $\boldsymbol{\nu} = (\nu_1, \nu_2, \ldots, \nu_d)$, then the eigenfunctions $\Phi_{\boldsymbol{\nu}}$ are

$$\Phi_{\boldsymbol{\nu}}(x_1, \ldots, x_d) = e^{2\pi i(\nu_1 x_1 + \ldots + \nu_d x_d)}, \quad \begin{array}{l} \nu_j = \ldots -1, 0, 1, \ldots \\ j = 1, 2, \ldots, d. \end{array}$$

Observing that

$$\begin{aligned} \Delta \Phi_{\boldsymbol{\nu}} &= [(2\pi\nu_1)^2 + \ldots + (2\pi\nu_d)^2] \quad \Phi_{\boldsymbol{\nu}} \\ &= \|2\pi\boldsymbol{\nu}\|^2 \quad \Phi_{\boldsymbol{\nu}} \end{aligned}$$

it is not hard to see that the $\sigma_{\boldsymbol{\nu}}^2$'s for the d-dimensional periodic stochastic process corresponding to the thin-plate spline penalty functional are

$$\sigma_{\boldsymbol{\nu}}^2(1) = \|2\pi\boldsymbol{\nu}\|^{-2m}$$

and the eigenvalues corresponding to the d-dimensional periodic version of kriging, given by the d-dimensional version of the covariance of (2.5.4) are, from (2.5.5),

$$\sigma_{\boldsymbol{\nu}}^2(2) = \sum_{l=m}^{m+k} \frac{\alpha_{l-m}}{\|2\pi\boldsymbol{\nu}\|^{2l}} = \frac{1}{\|2\pi\boldsymbol{\nu}\|^{2m}} \left(1 + \sum_{j=1}^{k} \frac{\alpha_j}{\|\boldsymbol{\nu}\|^{2j}}\right), \tag{3.2.1}$$

where we have set $\alpha_0 = 1$. Then

$$\sum_{\nu_1 \ldots \nu_d = -\infty}^{\infty} \left(1 - \frac{\sigma_{\boldsymbol{\nu}}^2(2)}{\sigma_{\boldsymbol{\nu}}^2(1)}\right)^2 = \sum_{\nu_1, \ldots \nu_d = -\infty}^{\infty} \left(\sum_{j=1}^{k} \frac{\alpha_j}{\|\boldsymbol{\nu}\|^{2j}}\right)^2 . \tag{3.2.2}$$

Looking at the lowest order term in (3.2.2)

$$\sum_{\nu_1, \ldots, \nu_\alpha = -\infty}^{\infty} \left(\frac{\alpha_1}{\|\boldsymbol{\nu}\|^2}\right)^2 = \alpha_1^2 \sum_{\nu_1, \ldots, \nu_d = -\infty}^{\infty} \frac{1}{\|\boldsymbol{\nu}\|^4} , \tag{3.2.3}$$

we estimate the sum by

$$\int \cdots \int_{\|\mathbf{x}\|>1} \frac{1}{\|\mathbf{x}\|^4} dx_1 \ldots dx_d = \int_{r>1} \cdots \int \frac{1}{r^4} r^{d-1} \, dr$$

where $\|\mathbf{x}\|^2 = x_1^2 + \ldots + x_d^2$ and the expression on the right is obtained by transforming to polar coordinates. The expression on the right will be finite if $4 - (d-1) = 5 - d > 1$. In particular, it will be finite for $d = 1, 2$, and 3. Thus the right-hand side of (3.2.3) will be finite for $d = 1, 2$, and 3, and the conclusion to be drawn is that P_1 and P_2 here are equivalent, and $\alpha_1, \ldots, \alpha_k$ cannot be estimated consistently. One can argue that the situation is analogous for the stochastic process with variogram

$$\tilde{E}(\tau) = |\tau|^{2m-1} + \alpha_1 |\tau|^{2m+1} + \ldots + \alpha_k |\tau|^{2m+2k-1}, \; (2m-1) > d \tag{3.2.4}$$

and if this is so, then $\alpha_1, \ldots, \alpha_k$ cannot be estimated consistently from data in a bounded region for $d = 1, 2$, and 3. Thus, in practice, if prior information is not available concerning $\alpha_1, \ldots, \alpha_k$, one might as well set $\alpha_1, \ldots, \alpha_k$ to zero, that is, use the thin-plate spline. I made observations to this effect in Wahba (1981b). In an elegant series of papers Stein (1985, 1987a, 1987b) has obtained results that imply the same thing.

CHAPTER 4

Estimating the Smoothing Parameter

4.1 The importance of a good choice of λ.

Figures 4.1, 4.2, and 4.3 from Wahba and Wold (1975) were part of the results of a Monte Carlo study to examine the behavior of ordinary cross validation (OCV) for estimating the smoothing parameter in a cubic smoothing spline. The dashed line in each of these figures is a plot of $f(x) = 4.26(e^{-x} - 4e^{-2x} + 3e^{-3x})$, and the dots enclosed in boxes represent values of

$$y_i = f\left(\frac{i}{n}\right) + \epsilon_i, \qquad i = 1, \ldots, 100, \tag{4.1.1}$$

where the ϵ_i's come from a random number generator simulating independently and identically distributed $\mathcal{N}(0, \sigma^2)$ random variables, with standard deviation $\sigma = .2$. The solid line in each figure is f_λ, the minimizer of $1/n \sum_{i=1}^n (y_i - f(i/n))^2 + \lambda \int_0^1 (f''(x))^2 dx$. In Figure 4.1 λ is too small, in Figure 4.2 too big, and in Figure 4.3 "about right." The parameter λ in Figure 4.3 was estimated by OCV, also known as the "leaving-out-one" method, to be described. Evidently the visual appearance of the picture is quite dependent on λ, which is not surprising as we recall that as λ runs from zero to ∞, f_λ runs from an interpolant to the data, to the straight line best fitting the data in a least squares sense. Figures 4.4–4.8 provide a two-dimensional example. Figure 4.4 gives a plot of a test function used by Franke (1979), this function is a linear combination of four normal density functions. Figure 4.5 gives a schematic plot of the data $y_i = f(x_1(i), x_2(i)) + \epsilon_i$, where the ϵ_i come from a random number generator as before. The (x_1, x_2) take on values on a 7×7 regular grid and the $n = 49$ y_i's have been joined by straight lines in an attempt to make the picture clearer. Figure 4.6 gives a plot of f_λ, the minimizer of

$$\frac{1}{n}\sum_{i=1}^{n}(y_i - f(x_1(i), x_2(i)))^2 + \iint_{-\infty}^{\infty} (f_{x_1x_1}^2 + 2f_{x_1x_2}^2 + f_{x_2x_2}^2)\, dx_1\, dx_2$$

with λ evidently too large, Figure 4.7 gives f_λ with λ too small, and Figure 4.8 gives f_λ with λ "about right." Figures 4.4–4.8 are from Wahba (1979c). Generalized cross validation (to be discussed) was used to choose λ in Figure 4.8.

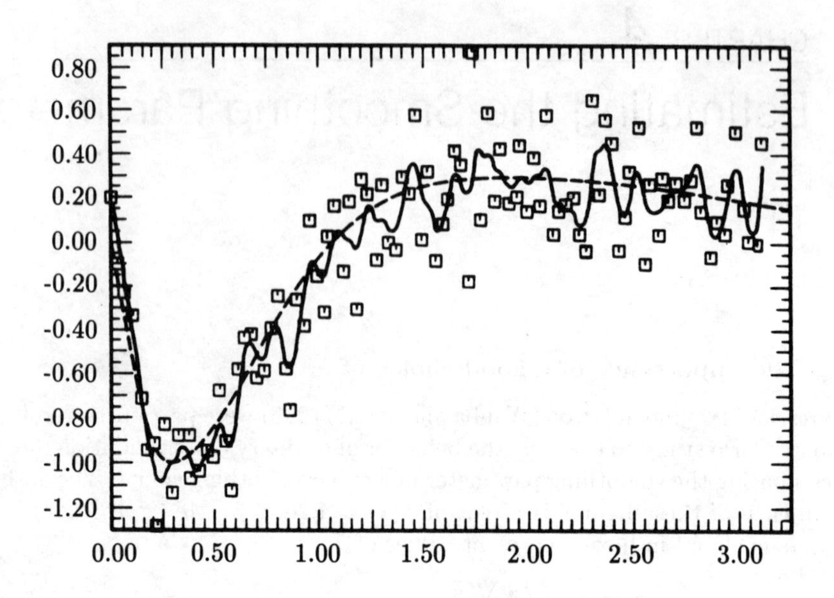

FIG. 4.1. *Data generated according to the model* (4.1.1). *Dashed curve is* $f(x)$. *Solid curve is fitted spline with* λ *too small.*

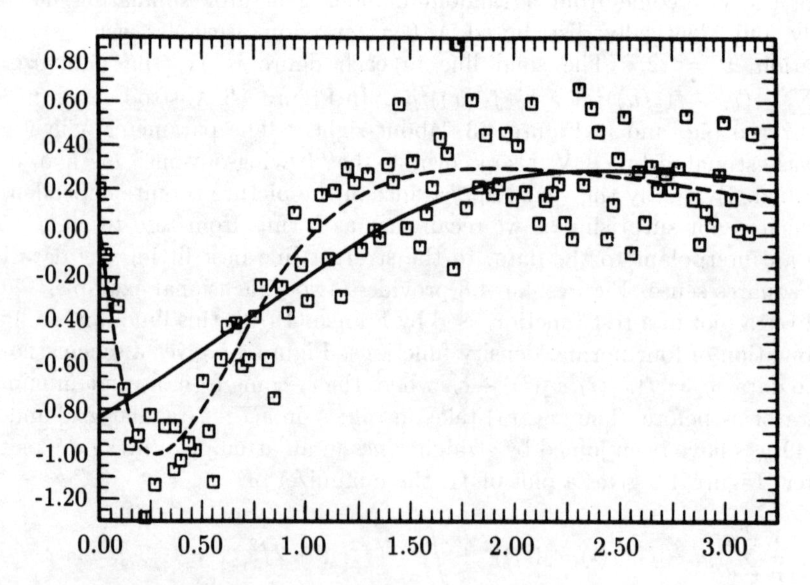

FIG. 4.2. *Same data as in Figure* 4.1. *Spline (solid curve) is fitted with* λ *too big.*

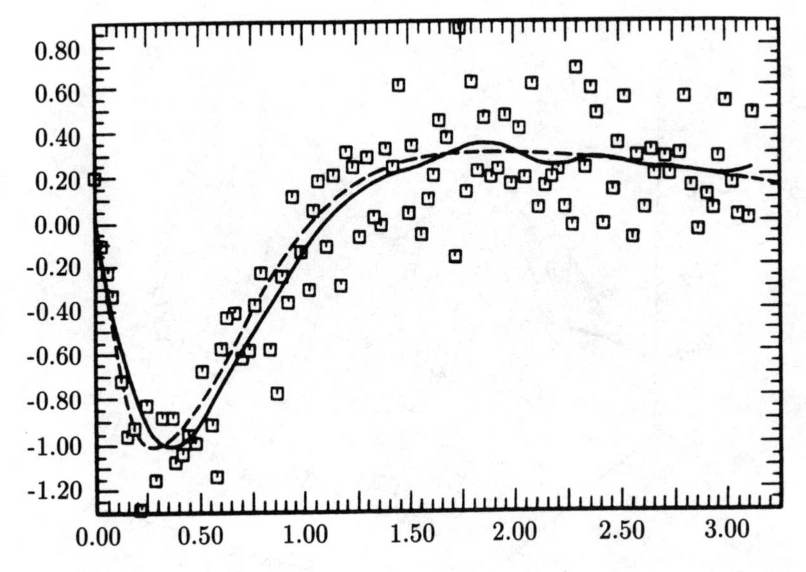

FIG. 4.3. *Same data as in Figure 4.2. Spline (solid curve) is fitted with the OCV estimate of* λ.

4.2 Ordinary cross validation and the "leaving-out-one" lemma.

Next we will explain these methods. Ordinary cross validation (OCV) goes as follows. Let $f_\lambda^{[k]}$ be the minimizer of

$$\frac{1}{n}\sum_{\substack{i=1\\ i\neq k}}^{n}(y_i - f(x_i))^2 + \lambda\int_0^1 (f''(u))^2\,du. \tag{4.2.1}$$

Then the "ordinary cross-validation function" $V_0(\lambda)$ is

$$V_0(\lambda) = \frac{1}{n}\sum_{k=1}^{n}\left(y_k - f_\lambda^{[k]}(x_k)\right)^2, \tag{4.2.2}$$

and the OCV estimate of λ is the minimizer of $V_0(\lambda)$. More generally, if we let $f_\lambda^{[k]}$ be the minimizer of

$$\frac{1}{n}\sum_{\substack{i=1\\ i\neq k}}^{n}(y_i - L_i f)^2 + \lambda\|P_1 f\|^2 \tag{4.2.3}$$

(assumed unique), then

$$V_0(\lambda) = \frac{1}{n}\sum_{k=1}^{n}(y_k - L_k f_\lambda^{[k]})^2. \tag{4.2.4}$$

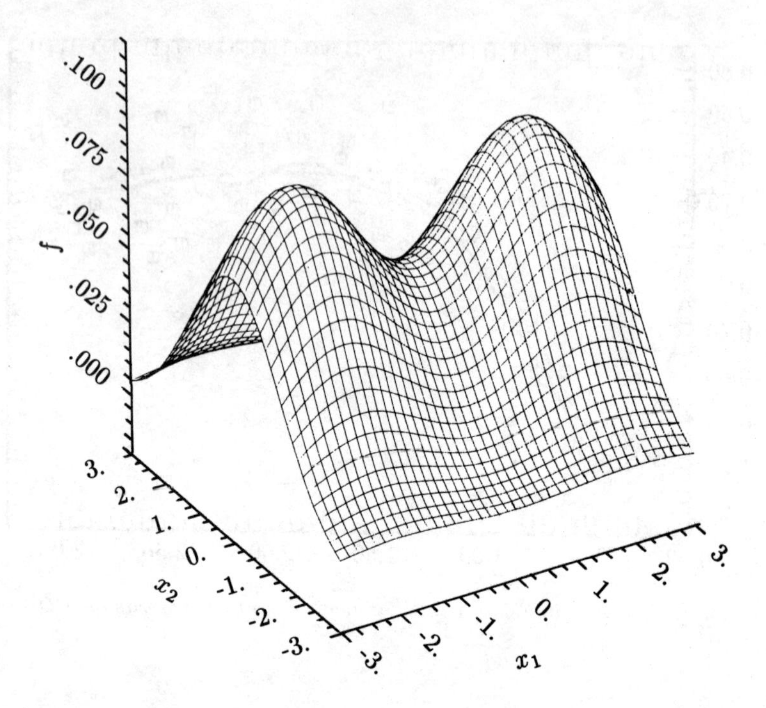

FIG. 4.4. *The actual surface.*

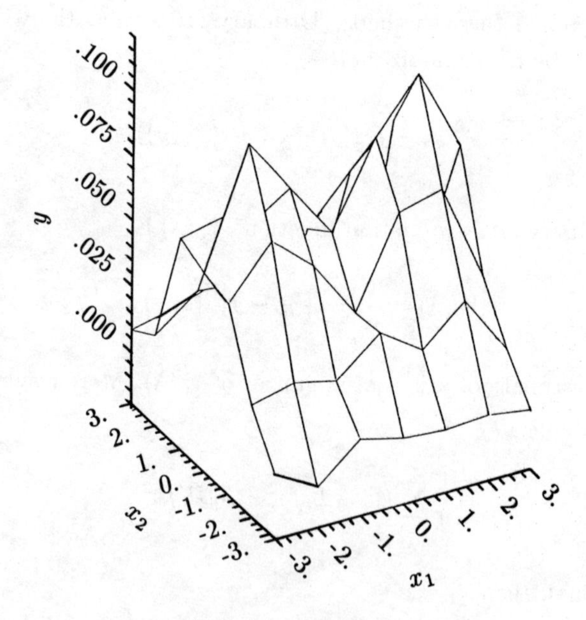

FIG. 4.5. *The data.*

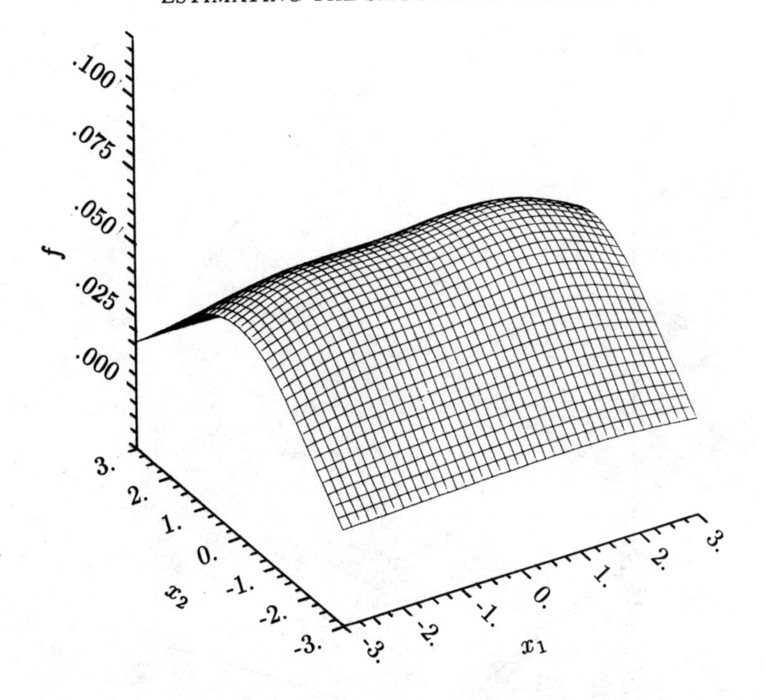

FIG. 4.6. f_λ *with* λ *too large,* $\lambda = 100\hat{\lambda}$.

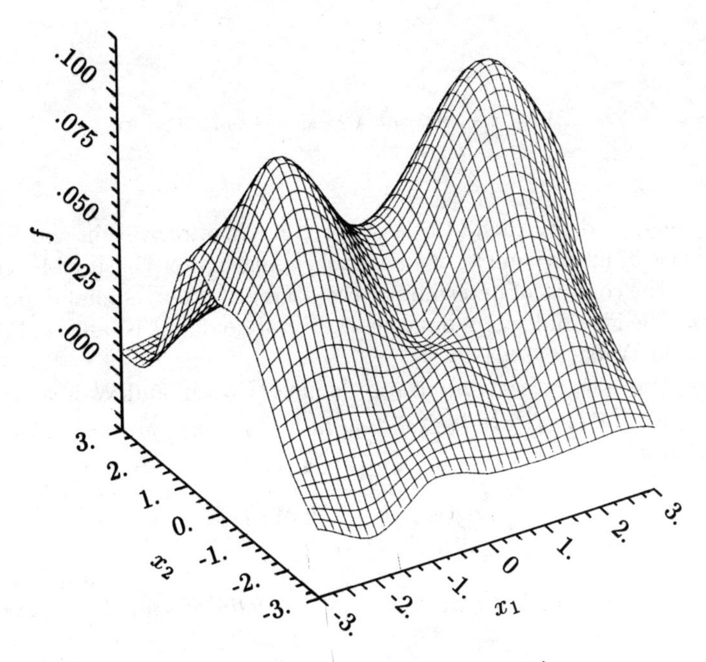

FIG. 4.7. f_λ *with* λ *too small,* $\lambda = .01\hat{\lambda}$.

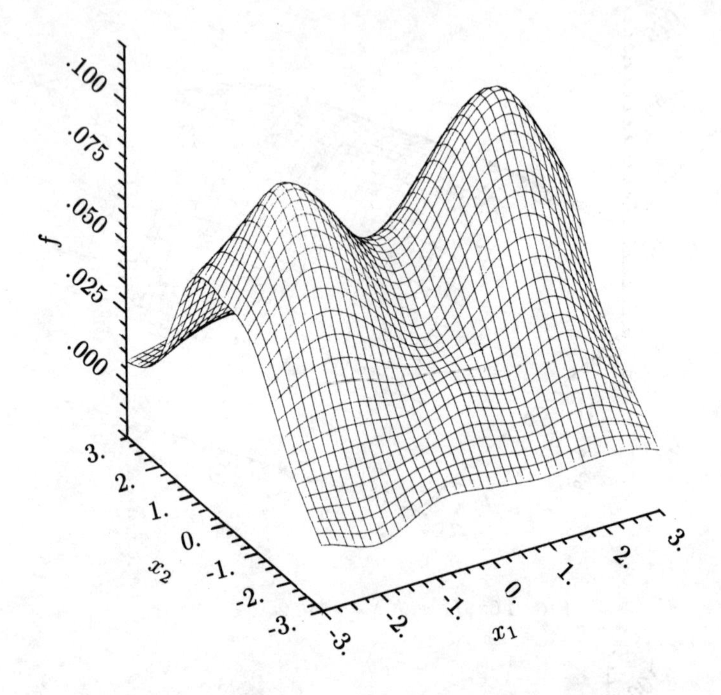

FIG. 4.8. *f_λ with λ estimated by GCV.*

OCV was suggested by Allen (1974) in the context of regression and by Wahba and Wold (1975) in the context of smoothing splines, after hearing Mervyn Stone discuss it in the context of determining the degree of a polynomial in polynomial regression. The idea of leaving out one or several no doubt is quite old (see, e.g., Mosteller and Wallace (1963)).

We now prove the "*leaving-out-one*" lemma (Craven and Wahba (1979)).

LEMMA 4.2.1. *Let $f_\lambda^{[k]}$ be the solution to the following problem. Find $f \in \mathcal{H}_R$ to minimize*

$$\frac{1}{n}\sum_{\substack{i=1\\ i\neq k}}^{n}(y_i - L_i f)^2 + \lambda\|P_1 f\|^2.$$

Fix k and z and let $h_\lambda[k,z]$ be the solution to the following problem. Find $f \in \mathcal{H}_R$ to minimize

$$\frac{1}{n}\left[(z - L_k f)^2 + \sum_{\substack{i=1\\ i\neq k}}^{n}(y_i - L_i f)^2\right] + \lambda\|P_1 f\|^2. \tag{4.2.5}$$

Then $h_\lambda[k, L_k f_\lambda^{[k]}] = f_\lambda^{[k]}$.

Proof. Let $\tilde{y}_k = L_k f_\lambda^{[k]}$, let $h = f_\lambda^{[k]}$, and let f be any element in $\mathcal{H}_R$ different from h. Then

$$\begin{aligned}
&\frac{1}{n}\left[(\tilde{y}_k - L_k h)^2 + \sum_{\substack{i=1\\ i\neq k}}^{n}(y_i - L_i h)^2\right] + \lambda\|P_1 h\|^2 \\
&= \frac{1}{n}\sum_{\substack{i=1\\ i\neq k}}^{n}(y_i - L_i h)^2 + \lambda\|P_1 h\|^2 \\
&< \frac{1}{n}\sum_{\substack{i=1\\ i\neq k}}^{n}(y_i - L_i f)^2 + \lambda\|P_1 f\|^2 \\
&\leq \frac{1}{n}\left[(\tilde{y}_k - L_k f)^2 + \sum_{\substack{i=1\\ i\neq k}}^{n}(y_i - L_i f)^2 + \lambda\|P_1 f\|^2\right]. \qquad (4.2.6)
\end{aligned}$$

Now consider the following identity:

$$y_k - L_k f_\lambda^{[k]} = \frac{(y_k - L_k f_\lambda)}{(1 - a_{kk}^*(\lambda))} \qquad (4.2.7)$$

where

$$a_{kk}^* = \frac{L_k f_\lambda - L_k f_\lambda^{[k]}}{y_k - L_k f_\lambda^{[k]}}. \qquad (4.2.8)$$

By the leaving-out-one lemma, and by letting $\tilde{y}_k = L_k f_\lambda^{[k]}$ and noting that $L_k f_\lambda = L_k h_\lambda[k, y_k]$ by definition, we can write

$$a_{kk}^* = \frac{L_k h_\lambda[k, y_k] - L_k h_\lambda[k, \tilde{y}_k]}{y_k - \tilde{y}_k}. \qquad (4.2.9)$$

Thus, looking at $L_k f_\lambda$ as a function of the kth data point, we see that $a_{kk}^*(\lambda)$ is nothing more than a divided difference of this function taken at y_k and $\tilde{y}_k$. However, $L_k f_\lambda$ is linear in each data point, so we can replace this divided difference by a derivative. Thus, we have shown that

$$a_{kk}^*(\lambda) = \frac{\partial L_k f_\lambda}{\partial y_k} = a_{kk}(\lambda), \qquad (4.2.10)$$

where $a_{kk}(\lambda)$ is the kkth entry of the influence matrix $A(\lambda)$, given in (1.3.23).

Thus, we have the following OCV *identity.*

THEOREM 4.2.1.

$$\frac{1}{n}\sum_{k=1}^{n}(y_k - L_k f_\lambda^{[k]})^2 \equiv V_0(\lambda) = \frac{1}{n}\sum_{k=1}^{n}(y_k - L_k f_\lambda)^2/(1 - a_{kk}(\lambda))^2. \qquad (4.2.11)$$

Later, we will generalize the optimization problem of (1.3.4) as follows. Find $f \in \mathcal{C} \subset \mathcal{H}_R$ to minimize

$$\frac{1}{n}\sum_{i=1}^{n}(y_i - N_i f)^2 + \lambda \|P_1 f\|^2 \tag{4.2.12}$$

where $\mathcal{C}$ is some closed convex set in $\mathcal{H}_R$, and $N_i f$ is a (possibly) nonlinear functional. Suppose that the N_i and $\mathcal{C}$ are such that (4.2.12) has a unique minimizer in $\mathcal{C}$, as does (4.2.12) with the kth term deleted. Then it is easy to see that the inequalities of (4.2.6) still hold, with L_i replaced by N_i. Therefore, if $f_\lambda^{[k]}$ is the minimizer in $\mathcal{C}$ of (4.2.12) with the kth term deleted, and $h_\lambda[k,z]$ is the minimizer in $\mathcal{C}$ of (4.2.12) with y_k replaced by z, then, as before, $h_\lambda[k, N_k f_\lambda^{[k]}] = f_\lambda^{[k]}$. Thus the OCV identity generalizes to

$$\frac{1}{n}\sum_{k=1}^{n}(y_k - N_k f_\lambda^{[k]})^2 = \frac{1}{n}\sum_{k=1}^{n}(y_k - N_k f_\lambda)^2/(1 - a_{kk}^*(\lambda))^2 \tag{4.2.13}$$

where

$$a_{kk}^* = \frac{N_k h_\lambda[k, y_k] - N_k h_\lambda[k, \tilde{y}_k]}{y_k - \tilde{y}_k} \tag{4.2.14}$$

with $\tilde{y}_k = N_k f_\lambda^{[k]}$. Now, however, $(\partial N_k f_\lambda / \partial y_k)|_{y_k}$, if it exists, will only be, in general, an approximation to $a_{kk}^*(\lambda)$ of (4.2.14).

4.3 Generalized cross validation.

Generalized cross validation (GCV) for the problem of (1.3.4) is obtained by replacing $a_{kk}(\lambda)$ by $\mu_1(\lambda) = 1/n \sum_{i=1}^{n} a_{ii}(\lambda) = 1/n \operatorname{Tr} A(\lambda)$. The GCV function $V(\lambda)$ is defined by

$$\begin{aligned} V(\lambda) &= \frac{1}{n}\sum_{k=1}^{n}(y_k - L_k f_\lambda)^2/(1 - \mu_1(\lambda))^2 \\ &\equiv \frac{1}{n}\|(I - A(\lambda))y\|^2 \Big/ \left[\frac{1}{n}\operatorname{Tr}(I - A(\lambda))\right]^2 . \end{aligned} \tag{4.3.1}$$

$V(\lambda)$ may be viewed as a weighted version of $V_0(\lambda)$, since

$$V(\lambda) = \frac{1}{n}\sum_{k=1}^{n}\left(y_k - L_k f_\lambda^{[k]}\right)^2 w_{kk}(\lambda)$$

where $w_{kk}(\lambda) = (1 - a_{kk}(\lambda))^2/(1 - \mu_1(\lambda))^2$. If $a_{kk}(\lambda)$ is independent of k, then $V_0(\lambda) \equiv V(\lambda)$. The "generalized" version was an attempt to achieve certain desirable invariance properties that do not generally hold for ordinary cross validation. Let Γ be any $n \times n$ orthogonal matrix, and consider a new data vector $\tilde{y} = \Gamma y$ and a new set of bounded linear functionals $(\tilde{L}_1, \ldots, \tilde{L}_n)' = \Gamma(L_1, \ldots, L_n)'$. The problem of estimating f from data

$$\tilde{y}_i = \tilde{L}_i f + \tilde{\epsilon}_i\ ,\ i = 1, \ldots, n,$$

where $\tilde{\epsilon} = \Gamma\epsilon$ is the same as the problem of estimating f from

$$y_i = L_i f + \epsilon_i, \; i = 1, \ldots, n,$$

since $\tilde{\epsilon} \sim \mathcal{N}(0, \sigma^2 I)$. However, it is not hard to see that, in general, OCV can give a different value of λ. The GCV estimate of λ is invariant under this transformation.

The original argument by which GCV was obtained from OCV can be described most easily with regards to a ridge regression problem (see Golub, Heath, and Wahba (1979)). Let

$$y = X\beta + \epsilon, \tag{4.3.2}$$

where $\epsilon \sim \mathcal{N}(0, \sigma^2 I)$, and to avoid irrelevant discussion suppose X is $n \times n$. β will be estimated as the minimizer of

$$\frac{1}{n}\|y - X\beta\|^2 + \lambda\, \beta'\beta.$$

Let the singular value decomposition (see Dongarra et al. (1979)) of X be UDV', and write

$$\tilde{y} = D\gamma + \tilde{\epsilon} \tag{4.3.3}$$

where $\tilde{y} = U'y$, $\gamma = V'\beta$, and $\tilde{\epsilon} = U'\epsilon$. The problem is invariant under this transformation. On the other hand, since

$$\tilde{y}_i = d_i \gamma_i + \tilde{\epsilon}_i \;,\; i = 1, \ldots, n, \tag{4.3.4}$$

where d_i is the iith entry of D, it is fairly clear that a leaving-out-one method of choosing λ is not going to work too well since the rows are uncoupled. (In fact, $V_0(\lambda)$ is independent of λ!) However, if the singular values of D come in pairs, there is an orthogonal matrix W for which WDW' is a symmetric circulant matrix, and symmetric circulant matrices may be viewed as having rows that are maximally coupled. Recall that a circulant matrix has the property that if the first row is $(\theta_0, \theta_1, \ldots, \theta_{n-1})$, then the jth row is $(\theta_{n-j}, \ldots, \theta_{n-1}, \theta_0, \theta_1, \ldots, \theta_{n-j-1})$. The matrix W is the discrete Fourier transform matrix and W and D are given, for even n, in Table 4.1. Transforming (4.3.3) by W gives

$$z = WDW'\delta + \xi \tag{4.3.5}$$

where $z = W\tilde{y}$, $\delta = W\gamma$, $\xi = W\epsilon$, and WDW' is circulant. Intuitively, the design matrix D has "maximally uncoupled" rows, while the design matrix WDW', being circulant, has "maximally coupled" rows. The influence matrix $A(\lambda)$ for the problem (4.3.5) is circulant and hence constant down the diagonal. GCV is equivalent here to transforming the original problem (4.3.2) into the "maximally coupled" form (4.3.5), doing OCV, and transforming back.

The GCV estimate $\hat{\lambda}$ of λ is known to have a number of favorable properties, both from practical experience and theoretically. For some Monte Carlo experimental results, see, e.g., Craven and Wahba (1979), Merz (1980), Nychka

TABLE 4.1
The discrete Fourier transform matrix W and corresponding diagonal matrix D.

$$W = \begin{pmatrix} -c_0 \\ -\sqrt{2}c_1 \\ \vdots \\ -\sqrt{2}c_{n/2-1} \\ -c_{n/2} \\ -\sqrt{2}s_1 \\ \vdots \\ -\sqrt{2}s_{n/2-1} \end{pmatrix}$$

where

$$\begin{aligned} c_0 &= \frac{1}{\sqrt{n}}(1,\ldots,1), \\ c_\nu &= \frac{1}{\sqrt{n}}(\cos 2\pi\nu\frac{1}{n}, \cos 2\pi\nu\frac{2}{n},\ldots,\cos 2\pi\nu\frac{n}{n}), \\ s_\nu &= \frac{1}{\sqrt{n}}(\sin 2\pi\nu\frac{1}{n}, \sin 2\pi\nu\frac{2}{n},\ldots,\sin 2\pi\nu\frac{n}{n}). \end{aligned}$$

$$D = \begin{pmatrix} d_0 & & & & & & & 0 \\ & d_1 & & & & & & \\ & & \ddots & & & & & \\ & & & d_{n/2-1} & & & & \\ & & & & \frac{1}{2}d_{n/2} & & & \\ & & & & & d_1 & & \\ & & & & & & \ddots & \\ 0 & & & & & & & d_{n/2-1} \end{pmatrix}$$

et al. (1984), Vogel (1986), Shahrary and Anderson (1989), Scott and Terrell (1987), Woltring (1985), the rejoinder in Hardle, Hall, and Marron (1988), etc. The so-called "weak cross-validation theorem" was proposed and nearly proved for the smoothing spline case in Craven and Wahba (1979). Utreras (1978, 1981b) completed the proof by obtaining rigorously certain properties of some eigenvalues necessary to complete the proof. Properties of eigenvalues in other cases were obtained by Utreras (1979, 1981b), Cox (1983), and others. See also Wahba (1977a). Strong theorems were obtained by Speckman (1985), Li(1985, 1986, 1987). Generalizations are discussed in Hurvich (1985), O'Sullivan (1986a), Altman (1987), Gu (1989b), Friedman and Silverman (1989). The arguments below are adapted from Craven and Wahba (1979) and Wahba (1985e).

4.4 Properties of the GCV Estimate of λ.

GCV is a *predictive mean-square error* criteria, which is not surprising given its source. Define the predictive mean-square error $T(\lambda)$ as

$$T(\lambda) = \frac{1}{n}\sum_{i=1}^{n}(L_i f_\lambda - L_i f)^2. \tag{4.4.1}$$

The GCV estimate of λ is an estimate of the minimizer of $T(\lambda)$. $T(\lambda)$ depends on the unknown f as well as the unknown $\epsilon_1, \ldots, \epsilon_n$. The expected value of $T(\lambda)$, $ET(\lambda)$ is given by

$$ET(\lambda) = E\frac{1}{n}\sum_{k=1}^{n}(L_k f_\lambda - L_k f)^2.$$

Letting $g = (L_1 f, \ldots, L_n f)'$ we have $(L_1 f_\lambda, \cdots, L_n f_\lambda)' = A(\lambda)y = A(\lambda)(g+\epsilon)$, and

$$\begin{aligned} ET(\lambda) &= \frac{1}{n}E\|A(\lambda)(g+\epsilon) - g\|^2 \\ &= \frac{1}{n}\|(I - A(\lambda))g\|^2 + \frac{\sigma^2}{n}\,\mathrm{Tr}\; A^2(\lambda) \\ &= b^2(\lambda) + \sigma^2\mu_2(\lambda), \;\text{ say.} \end{aligned}$$

These terms are known as the bias and variance terms, respectively. Using the representation for $I - A(\lambda)$ given in (1.3.23),

$$I - A(\lambda) = n\lambda Q_2(Q_2'(\Sigma + n\lambda I)Q_2)^{-1}Q_2',$$

letting the eigenvector eigenvalue decomposition of $Q_2'\Sigma Q_2$ be UDU', where U is $(n-M)\times(n-M)$ orthogonal and D is diagonal and $\Gamma = Q_2\,U$, we have

$$I - A(\lambda) = n\lambda\Gamma(D + n\lambda I)^{-1}\Gamma'. \tag{4.4.2}$$

Letting

$$h = \Gamma' g$$

we have

$$
\begin{aligned}
b^2(\lambda) &= \frac{1}{n}\sum_{\nu=1}^{n-M}\left(\frac{n\lambda h_{\nu n}}{\lambda_{\nu n}+n\lambda}\right)^2, \\
\mu_2(\lambda) &= \frac{1}{n}\left(\sum_{\nu=1}^{n-M}\left(\frac{\lambda_{\nu n}}{\lambda_{\nu n}+n\lambda}\right)^2 + M\right)
\end{aligned}
\tag{4.4.3}
$$

where $h_{\nu n}$, $\nu = 1, \ldots, n-M$ are the components of h, and $\lambda_{\nu n}$ are the diagonal entries of D. If f is in the null space of P_1, that is, f is of the form

$$f = \sum_{\nu=1}^{M} \theta_\nu \phi_\nu,$$

then $g = (L_1 f, \ldots, L_n f)' = T\theta$, where $\theta = (\theta_1, \ldots, \theta_M)'$, and then $h = \Gamma' g = U'Q_2' g = U'Q_2' T\theta = 0$, by the construction of Q_2 in (1.3.18). Then $\mu_2(\lambda)$ is a monotone decreasing function of λ and is minimized for $\lambda = \infty$, which corresponds to f_∞ being the least squares regression of the data on span $\{\phi_1, \ldots, \phi_M\}$, with $\mu_2(\infty) = M/n$. If $\sum_{\nu=1}^{n-M} h_{\nu n}^2 > 0$, then $b^2(\lambda)$ is a monotone increasing function of λ, with $(d/d\lambda) b^2(\lambda)|_{\lambda=0} = 0$, while $\mu_2(\lambda)$ is a monotone decreasing function of λ with strictly negative derivative at $\lambda = 0$, so that $ET(\lambda)$ will have (at least) one minimizer $\lambda^* > 0$.

The "weak GCV theorem" says that there exists a sequence (as $n \to 0$) of minimizers $\tilde{\lambda}$ of $EV(\lambda)$ that comes close to achieving the minimum value of $\min_\lambda ET(\lambda)$. That is, let the expectation inefficiency I^* be defined by

$$I^* = \frac{ET(\tilde{\lambda})}{ET(\lambda^*)}.$$

Then, under some general circumstances to be discussed, $I^* \downarrow 1$ as $n \to \infty$.

We will outline the argument. First,

$$EV(\lambda) = \frac{b^2(\lambda) + \sigma^2(1 - 2\mu_1(\lambda) + \mu_2(\lambda))}{(1-\mu_1(\lambda))^2}, \tag{4.4.4}$$

where

$$\mu_1(\lambda) = \frac{1}{n}\left[\sum_{\nu=1}^{n-M}\frac{\lambda_{\nu n}}{\lambda_{\nu n}+n\lambda} + M\right]. \tag{4.4.5}$$

As before, if $\|P_1 f\|^2 = 0$, then $b^2(\lambda) = 0$, and

$$EV(\lambda) = \frac{\sigma^2}{n}\sum_{\nu=1}^{n-M}\left(\frac{n\lambda}{\lambda_{\nu n}+n\lambda}\right)^2 \Big/ \left(\frac{1}{n}\sum_{\nu=1}^{n-M}\frac{n\lambda}{\lambda_{\nu n}+n\lambda}\right)^2, \tag{4.4.6}$$

which is minimized for $\lambda = \infty$, the same as for $ET(\lambda)$, so in this case $I^* = 1$.

We now proceed to the general case. First, some algebraic manipulations give

$$\frac{ET(\lambda)-(EV(\lambda)-\sigma^2)}{ET(\lambda)}=\frac{-\mu_1(2-\mu_1)}{(1-\mu_1)^2}+\frac{\sigma^2}{b^2+\sigma^2\mu_2}\cdot\frac{\mu_1^2}{(1-\mu_1)^2}$$

and so

$$\frac{|ET(\lambda)-(EV(\lambda)-\sigma^2)|}{ET(\lambda)}\le h(\lambda) \tag{4.4.7}$$

where

$$h(\lambda)=\left[2\mu_1(\lambda)+\frac{\mu_1^2(\lambda)}{\mu_2(\lambda)}\right]\frac{1}{(1-\mu_1(\lambda))^2}. \tag{4.4.8}$$

Now, using the fact that $\mu_2(\lambda)\ge\mu_1^2(\lambda)$, for all λ, it follows that $EV(\lambda)\ge\sigma^2$, so that (4.4.7) gives

$$ET(\lambda)(1-h(\lambda))\le EV(\lambda)-\sigma^2\le ET(\lambda)(1+h(\lambda)),\ \text{for all } \lambda. \tag{4.4.9}$$

Letting λ^* be the minimizer of $ET(\lambda)$, we obtain

$$EV(\lambda^*)-\sigma^2\le ET(\lambda^*)(1+h(\lambda^*))$$

and there must be (at least) one minimizer $\tilde{\lambda}$ of $EV(\lambda)$ in the nonempty set $\Lambda=\{\lambda: EV(\lambda)-\sigma^2\le EV(\lambda^*)-\sigma^2\}$ (see Figure 4.9). Thus

$$ET(\tilde{\lambda})(1-h(\tilde{\lambda}))\le EV(\tilde{\lambda})-\sigma^2\le EV(\lambda^*)-\sigma^2\le ET(\lambda^*)(1+h(\lambda^*)). \tag{4.4.10}$$

Then (provided $h(\tilde{\lambda})<1$),

$$\frac{ET(\tilde{\lambda})}{ET(\lambda^*)}\le\frac{1+h(\lambda^*)}{1-h(\tilde{\lambda})}, \tag{4.4.11}$$

and, if $h(\lambda^*)$ and $h(\tilde{\lambda})\to 0$, then

$$\frac{ET(\tilde{\lambda})}{ET(\lambda^*)}\downarrow 1. \tag{4.4.12}$$

$h(\lambda^*)$ and $h(\tilde{\lambda})$ will tend to zero if $\mu_1(\lambda^*),\mu_1(\tilde{\lambda}),\mu_1^2(\lambda^*)/(\mu_2(\lambda^*))$ and $\mu_1^2(\tilde{\lambda})/(\mu_2(\tilde{\lambda}))$ tend to zero.

In many interesting cases the eigenvalues $\lambda_{\nu n}$ behave roughly as do $n\nu^{-q}$ for some real number $q>1$, and the expressions

$$\begin{aligned}\mu_\tau(\lambda) &= \frac{1}{n}\sum_{\nu=1}^{n-M}\left(\frac{\lambda_{\nu n}}{\lambda_{\nu n}+n\lambda}\right)^\tau\simeq\frac{1}{n}\sum\frac{1}{(1+\lambda\nu^q)^\tau},\ \tau=1,2\\ &\simeq \frac{1}{n}\int\frac{1}{(1+\lambda x^q)^\tau}\simeq\frac{c_{\tau q}}{n\lambda^{1/q}},\quad \tau=1,2,\quad q>1\end{aligned}$$

are valid to the accuracy needed in the proofs.

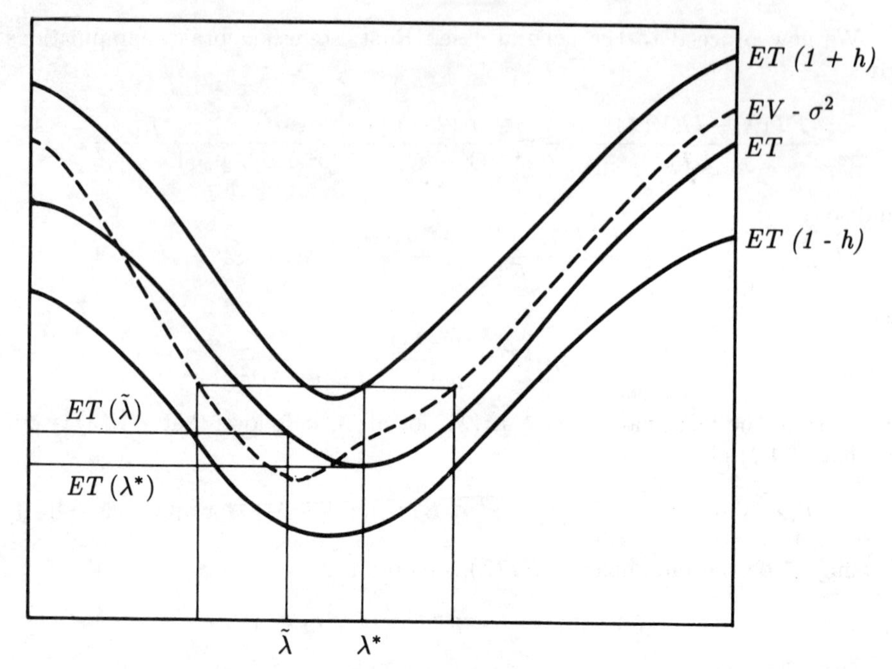

FIG. 4.9. *Graphical suggestion of the proof of the weak* GCV *theorem.*

We only give the reader a very crude argument in support of (4.4.12) and refer the reader to Cox (1988), Utreras (1978) for more rigorous results. The crude argument, for roughly equally spaced evaluation functionals, goes as follows. Suppose

$$R(s,t) = \sum_{\nu=1}^{\infty} \lambda_\nu \Phi_\nu(s)\Phi_\nu(t),$$

consider the matrix $\sum$ with ijth entry

$$\begin{aligned} R(t_i,t_j) &= \sum_{\nu=1}^{\infty} \lambda_\nu \Phi_\nu(t_i)\Phi_\nu(t_j) \\ &\approx \sum_{\nu=1}^{n} n\lambda_\nu \frac{\Phi_\nu(t_i)}{\sqrt{n}} \frac{\Phi_\nu(t_j)}{\sqrt{n}}. \end{aligned}$$

If

$$\frac{1}{n}\sum_{l=1}^{n} \Phi_\nu(t_l)\Phi_\mu(t_l) \simeq \int \Phi_\nu(s)\Phi_\mu(s)ds = \delta_{\mu,\nu} \tag{4.4.13}$$

then roughly $(1/\sqrt{n}\Phi_\nu(t_1),\ldots,1/\sqrt{n}\Phi_\nu(t_n))'$, $\nu = 1,\ldots,n$ are the eigenvectors of Σ and (again roughly), $n\lambda_\nu$, $\nu = 1,2,\ldots$ are the eigenvalues of Σ. The asymptotic behavior of the eigenvalues of $Q_2'\Sigma Q_2$ does not differ "much" from the asymptotic behavior of the eigenvalues of Σ. In particular, if $\alpha_1 \geq \ldots \geq \alpha_n$ are the eigenvalues of Σ and $\lambda_{1n},\ldots,\lambda_{n-M,n}$ are the eigenvalues of $Q_2'\Sigma Q_2$, then, by the variational definition of eigenvalues,

$$\alpha_1 \geq \lambda_{1,n} \geq \alpha_{M+1}$$

$$\alpha_2 \geq \lambda_{2,n} \geq \alpha_{M+2}$$

$$\vdots$$

$$\alpha_{n-M} \geq \lambda_{n-M,n} \geq \alpha_n.$$

For the reproducing kernel of W_m^0 (per) of (2.1.4) it is easy to see that (4.4.13) holds (exactly, for $t_l = l/n$), and $\lambda_{\nu n} \simeq n(2\pi\nu)^{-2m}$. If R is a Green's function for a linear differential operator, then the eigenvalues of R can be expected to behave as do the inverses of the eigenvalues of the linear differential operator (see Naimark (1967)).

To study $b^2(\lambda)$, we have the lemma:

$$b^2(\lambda) \leq \lambda \|P_1 f\|^2. \tag{4.4.14}$$

The proof follows upon letting $g = (L_1 f, \ldots, L_n f)'$ and noting that $A(\lambda)g = (L_1 f_\lambda^*, \ldots, L_n f_\lambda^*)'$, where f_λ^* is the solution to the following problem. Find $h \in \mathcal{H}_R$ to minimize

$$\frac{1}{n}\sum_{i=1}^{n}(g_i - L_i h)^2 + \lambda\|P_1 h\|^2.$$

Then

$$\begin{aligned}
&\frac{1}{n}\|(I - A(\lambda))g\|^2 + \lambda\|P_1 f_\lambda^*\|^2 \\
&= \frac{1}{n}\sum_{i=1}^{n}(g_i - L_i f_\lambda^*)^2 + \lambda\|P_1 f_\lambda^*\|^2 \\
&\leq \frac{1}{n}\sum_{i=1}^{n}(g_i - L_i f)^2 + \lambda\|P_1 f\|^2 \\
&= \lambda\|P_1 f\|^2.
\end{aligned}$$

If $\mu_2(\lambda) \simeq O(1/n\lambda^{1/q})$, then

$$ET(\lambda) \leq O(\lambda) + O\left(\frac{1}{n\lambda^{1/q}}\right) \tag{4.4.15}$$

and thus $ET(\lambda) \downarrow 0$ provided $\lambda \to 0$ and $n\lambda^{1/q} \to \infty$. Furthermore, it can be argued that if λ does not tend to zero (and of course if $n\lambda^{1/q}$ does not tend to infinity), then $ET(\lambda)$ cannot tend to zero. Thus $\mu_T(\lambda^*) \to 0$. Now $EV(\tilde{\lambda}) \downarrow \sigma^2$, since $EV(\tilde{\lambda}) - \sigma^2 \leq ET(\lambda^*)(1 + h(\lambda^*)) \to 0$. If $\sum h_{\nu n}^2 > 0$ it is necessary that $\tilde{\lambda} \to 0$, $n\tilde{\lambda}^{1/q} \to \infty$ in order that $EV(\tilde{\lambda}) \downarrow \sigma^2$ so that the following can be concluded:

$$I^* \downarrow 1.$$

Figure 4.10 gives a plot of $T(\lambda)$ and $V(\lambda)$ for the test function and experimental data of Figures 4.4 and 4.5. It can be seen that $V(\lambda)$ roughly behaves as does $T(\lambda) + \sigma^2$ in the neighborhood of the minimizer of $T(\lambda)$. The

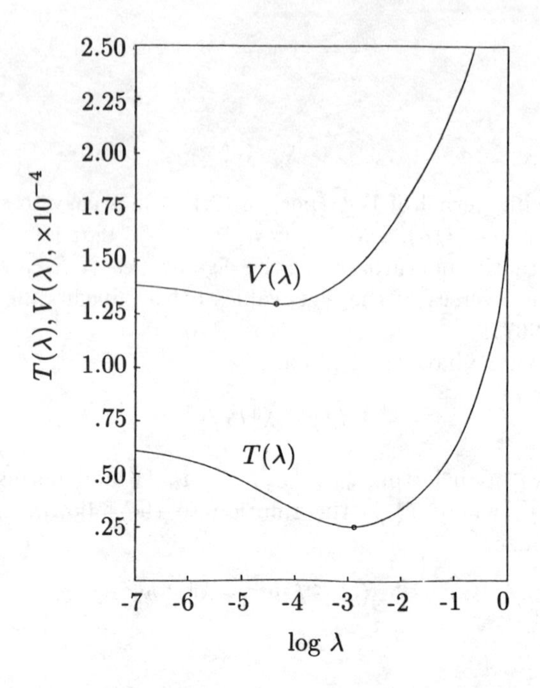

FIG. 4.10. *The mean square error* $T(\lambda)$ *and the cross validation function* $V(\lambda)$.

sample size was $n = 49$ here. This behavior of V relative to T generally becomes more striking in Monte Carlo experiments as n gets large.

We remark that the parameter m in the penalty functional for polynomial and thin-plate splines can also be estimated by GCV (see Wahba and Wendelberger (1980), Gamber (1979)). V is minimized for each fixed value of m and then the m with the smallest $V(\hat{\lambda})$ is selected.

4.5 Convergence rates with the optimal λ.

It can be seen from (4.4.15) that if $\mu_2(\lambda) = O(1/n\lambda^{1/q})$ and λ is taken as $O(1/n^{q/(q+1)})$, then $ET(\lambda^*) \le O(1/n^{q/(q+1)})$. If additional conditions hold on the $L_i f$ (that is, on the sequence $(L_{1n}f, \ldots, L_{nn}f)$, $n = 1, 2, \ldots$) then higher rates of convergence can be obtained.

Considering (4.4.3) for the bias, we have

$$\begin{aligned} b^2(\lambda) &= \frac{1}{n}\sum_{\nu=1}^{n-M}\left(\frac{n\lambda h_{\nu n}}{\lambda_{\nu n}+n\lambda}\right)^2 \\ &= \lambda^p \sum_{\nu=1}^{n-M}\left(\frac{n\lambda}{\lambda_{\nu n}+n\lambda}\right)^{2-p}\frac{(h_{\nu n}^2/n)}{(\lambda_{\nu n}/n+\lambda)^p} \\ &\le \lambda^p \sum_{\nu=1}^{n-M}\frac{(h_{\nu n}^2/n)}{(\lambda_{\nu n}/n)^p} \quad \text{for any } p \in [0,2]. \end{aligned} \tag{4.5.1}$$

If $\sum_{\nu=1}^{n-M}((h_{\nu n}^2/n)/(\lambda_{\nu n}/n)^p)$ is bounded as $n \to \infty$ for some p in $(1, 2]$, then

$$b^2(\lambda) \le O(\lambda^p) \tag{4.5.2}$$

as $\lambda \to 0$, and if $\mu_2(\lambda) = O(1/n\lambda^{1/q})$, then

$$ET(\lambda) \le O(\lambda^p) + O\left(\frac{1}{n\lambda^{1/q}}\right),$$

and upon taking $\lambda = O(1/n^{q/(pq+1)})$ we have

$$ET(\lambda^*) \le O\left(\frac{1}{n^{pq/(pq+1)}}\right). \tag{4.5.3}$$

We know from (4.4.14) that (4.5.2) always holds for $p = 1$; we show this fact another way, by proving that $\sum_{\nu=1}^{n-M}(h_{\nu n}^2/\lambda_{\nu n}) \le \|P_1 f\|^2$. Letting f_0 be that element in $\mathcal{H}_R$ that minimizes $\|P_1 f_0\|^2$ subject to

$$L_i f_0 = L_i f \equiv g_i, \text{ say,}$$

we have, using the calculations in Chapter 1, with $y_i = g_i$ and $\lambda = 0$, that

$$f_0 = \sum_{i=1}^{n} c_i^0 \xi_i + \sum_{\nu=1}^{M} d_\nu^0 \phi_\nu$$

where c^0 and d^0 satisfy $\Sigma c^0 + T d^0 = g$, $T' c^0 = 0$, and so $c^0 = Q_2(Q_2'\Sigma Q_2)^{-1}Q_2' g$. Now $\|P_1 f\|^2 \ge \|P_1 f_0\|^2 = c^{0'}\Sigma c^0 = g' Q_2(Q_2'\Sigma Q_2)^{-1}Q_2' g = \sum_{\nu=1}^{n-M} h_{\nu n}^2/\lambda_{\nu n}$.

An example to show when (4.5.1) holds for some $p > 1$ goes as follows. In the case $\mathcal{H}_R = W_m$ (per) with $L_i f = f(i/n)$, $f(t) = \sum f_\nu \Phi_\nu(t)$ (the Φ_ν are sines and cosines here), then $h_{\nu n} \approx \sqrt{n} f_\nu$ by the same argument as that leading up to (4.4.13), $\lambda_{\nu n} \simeq n(2\pi\nu)^{-2m}$, and if $\int_0^1 (f^{(pm)}(x))^2\, dx < \infty$, then

$$\infty > \sum_{\nu=1}^{\infty} (2\pi\nu)^{2pm} f_\nu^2 = \int_0^1 (f^{(pm)}(x))^2\, dx \simeq \sum_{\nu=1}^{n-m} \frac{h_{\nu n}^2/n}{\lambda_{\nu n}^{1+p}} (1 + o(1)).$$

Let

$$R^p(s,t) = \sum_{\nu=1}^{\infty} \lambda_\nu^p \Phi_\nu(s)\Phi_\nu(t)$$

for some $p \in (1, 2]$. $f \in \mathcal{H}_{R^p}$ if and only if

$$\sum_{\nu=1}^{\infty} \frac{f_\nu^2}{\lambda_\nu^p} < \infty \tag{4.5.4}$$

where

$$f_\nu = \int f(t)\Phi_\nu(t)\, dt.$$

A general argument similar to that surrounding (4.4.13), (4.5.2), and (4.5.3) would suggest that if the L_i's are roughly uniformly spaced evaluation functionals and $f \in \mathcal{H}_{R^p}$, and $\lambda_\nu = O(\nu^{-q})$, then convergence rates of $O(1/n^{pq/(pq+1)})$ are available. For more general L_i, see Wahba (1985e). Convergence rates for smoothing splines under various assumptions have been found by many authors. See, e.g., Davies and Anderssen (1985), Cox (1983, 1984, 1988), Craven and Wahba (1979), Johnstone and Silverman (1988), Lukas (1981), Ragozin (1983), Rice and Rosenblatt (1983), Silverman (1982), Speckman (1985), Utreras (1981b), Wahba (1977a), and Wahba and Wang (1987).

4.6 Other estimates of λ similar to GCV.

We remark that a variety of criteria $C(\lambda)$ have been proposed such that $\tilde{\lambda}$ is estimated as the minimizer of $C(\lambda)$, where $C(\lambda)$ is of the form

$$C(\lambda) = \|(I - A(\lambda))y\|^2 c(\lambda) \tag{4.6.1}$$

where $c(\lambda) = 1 + 2\mu_1(\lambda) + o(\mu_1(\lambda))$ when $\mu_1 \to 0$ (see Hardle, Hall, and Marron (1988)). Such estimates will have a sequence of minimizers that satisfy the weak GCV theorem.

Note that

$$V(\lambda) = \frac{1}{n}\sum_{\nu=1}^{n-M}\left(\frac{n\lambda}{\lambda_{\nu n} + n\lambda}\right)^2 z_{\nu n}^2 \Big/ \left(\frac{1}{n}\sum_{\nu=1}^{n-M}\frac{n\lambda}{\lambda_{\nu n} + n\lambda}\right)^2 \tag{4.6.2}$$

where $z_n = (z_{1n}, \ldots, z_{n-M,n})' = \Gamma' y$, where $\Gamma = Q_2 U$ as in (4.4.2). Provided the $\lambda_{\nu n}$ are nonzero,

$$\lim_{\lambda \to 0} V(\lambda) = \frac{1}{n}\sum \frac{z_{\nu n}}{\lambda_{\nu n}^2}^2 \Big/ \left(\frac{1}{n}\sum \frac{1}{\lambda_{\nu n}}\right)^2 > 0. \tag{4.6.3}$$

However, unless $c(\lambda)$ has a pole of order at least $1/\lambda^2$ as $\lambda \to 0$, then $C(\lambda)$ of (4.6.1) will be zero at zero, so that in practice the criterion is unsuitable. For n large, the $\lambda_{\nu n}$ may be very small, and the calculation of V or C in the obvious way may be unstable near zero; this fact has possibly masked the unsuitability of certain criteria of this form C in Monte Carlo studies.

4.7 More on other estimates.

When σ^2 is known, an unbiased risk estimate is available for λ. This type of estimate was suggested by Mallows (1973) in the regression case, and applied to spline smoothing by Craven and Wahba (1979) (see also Hudson (1974)). Recalling that $ET(\lambda) = (1/n)\|(I - A(\lambda))g\|^2 + (\sigma^2/n)\operatorname{Tr} A^2(\lambda)$, we let

$$\hat{T}(\lambda) = \frac{1}{n}\|(I - A(\lambda)y\|^2 - \frac{\sigma^2}{n}\operatorname{Tr}(I - A(\lambda))^2 + \frac{\sigma^2}{n}\operatorname{Tr} A^2(\lambda). \tag{4.7.1}$$

It is not hard to show that $ET(\lambda) = E\hat{T}(\lambda)$. The numerical experiments in Craven and Wahba show that the GCV estimate and the unbiased risk estimate behave essentially the same, to the accuracy of the experiment, when the same σ^2 is used in (4.7.1) as in generating the experimental data. It is probably true that a fairly good estimate of σ^2 would be required in practice to make this method work well. Several authors have suggested the so-called discrepancy method: Choose λ so that

$$\frac{1}{n}\|(I - A(\lambda))y\|^2 = \sigma^2. \tag{4.7.2}$$

The left-hand side is a monotone nondecreasing function of λ, and if $(1/n)\|(I - A(\infty))y\|^2$ (= the residual sum of squares after regression on the null space of $\|P_1(\cdot)\|^2$) is at least as large as σ^2, there will be a unique λ satisfying (4.7.2). We claim that this is not a very good estimate of the minimizer of $T(\lambda)$. Wahba (1975) showed that if λ^* is the minimizer of $ET(\lambda)$, then

$$E\frac{1}{n}\|(I - A(\lambda^*))y\|^2 = k\sigma^2(1 + o(1))$$

where k is a factor less than one. The experimental results in Craven and Wahba are consistent with these results, the discrepancy estimate λ_{dis} of λ being naturally larger than λ^* with $T(\lambda_{\text{dis}})/T(\lambda_{\text{opt}}) >> T(\lambda_{\text{GCV}})/T(\lambda_{\text{opt}})$, λ_{opt} being the minimizer of $T(\lambda)$.

By analogy with regression, I have suggested that $\text{Tr}\, A(\lambda)$ be called the degrees of freedom for signal when λ is used. (Note that $M \leq$ d.f. signal $\leq n$), and this suggests an estimate for σ^2, as

$$\hat{\sigma}^2 = \hat{\sigma}^2(\hat{\lambda}) = \frac{\|(I - A(\hat{\lambda}))y\|^2}{\text{Tr}\,(I - A(\hat{\lambda}))}$$

where $\hat{\lambda}$ is the GCV estimate of λ. Good numerical results for $\hat{\sigma}^2$ were obtained in Wahba (1983) although no theoretical properties of this estimate were given. Other estimates for σ^2 have been proposed (see, for example Buja, Hastie, and Tibshirani (1989). Hall and Titterington (1987) have proposed estimating λ as the solution to

$$\frac{\|(I - A(\lambda))y\|^2}{\text{Tr}\,(I - A(\lambda))} = \sigma^2$$

when σ^2 is known. It is not known how this estimate would compare, say, with the unbiased risk estimate.

4.8 The generalized maximum likelihood estimate of λ.

A maximum likelihood estimate of λ based on the Bayes model was suggested by Anderssen and Bloomfield (1974) in the case of a stationary time series, and by Wecker and Ansley (1983) in the smoothing spline case (see also Barry (1983)).

Beginning with the stochastic model, (1.5.8) gives

$$y \sim N(0, b(\eta TT' + \Sigma + n\lambda I)) \tag{4.8.1}$$

where $\eta = a/b$ and a, b, T, and λ are as in (1.5.9), and $\lambda = \sigma^2/nb$.

Let

$$\begin{pmatrix} z \\ \cdots \\ w \end{pmatrix} = \begin{pmatrix} Q_2' \\ \cdots \\ \frac{1}{\sqrt{\eta}}T' \end{pmatrix} y, \tag{4.8.2}$$

where $Q_2'T = 0$, as in (1.3.18). Then

$$z \sim N(0, b(Q_2'\Sigma Q_2 + n\lambda I)), \tag{4.8.3}$$

$$\lim_{\eta\to\infty} Ezw' = 0,$$

$$\lim_{\eta\to\infty} Eww' = b(T'T)(T'T).$$

It was argued in Wahba (1985e) that the maximum likelihood estimate of λ here should be based on (4.8.3) since the distribution of w is independent of λ. This estimate was called the GML estimate. A straightforward maximization of the likelihood of (4.8.3) with respect to b and λ gives the GML estimate of λ as the minimizer of

$$\begin{aligned} M(\lambda) &= \frac{z'(Q_2'\Sigma Q_2 + n\lambda I)^{-1}z}{[\det(Q_2'\Sigma Q_2 + n\lambda I)^{-1}]^{1/(n-M)}} \\ &= \frac{y'Q_2(Q_2'\Sigma Q_2 + n\lambda I)^{-1}Q_2'y}{[\det(Q_2'\Sigma Q_2 + n\lambda I)^{-1}]^{1/(n-M)}}. \end{aligned}$$

Multiplying the top and bottom of this expression by $n\lambda$ results in an expression that is readily compared with $V(\lambda)$, viz.

$$M(\lambda) = \frac{y'(I - A(\lambda))y}{[\det^{+}(I - A(\lambda))]^{1/(n-M)}} \tag{4.8.4}$$

where $\det^{+}$ is the product of the nonzero eigenvalues. We remark that Wecker and Ansley (1983) included the M components of w as unknown parameters in the likelihood function. After minimizing with respect to w, they got a (slightly) different equation for "the" maximum likelihood estimate of λ. (See O'Hagan (1976) for other examples of this phenomenon.) We also note that if either σ^2 or b were known then a different expression for the maximum likelihood estimate of λ would be obtained.

It is shown in Wahba (1985e) that if $\|P_1 f\|^2 > 0$ and $\sum_{\nu=1}^{n-M}(h_{\nu n}^2/n)/(\lambda_{\nu n}/n)^p$ is bounded as $n \to \infty$ and $\mu_1(\lambda)$ and $\mu_2(\lambda)$ are $O(1/n\lambda^{1/2q})$ for some $p \in (1,2]$, $q > 1$ then $(d/d\lambda)M(\lambda) = 0$ for $\lambda = \lambda_{\mathrm{GML}} = O(1/n^{q/(q+1)})$, independent of p. Thus, asymptotically, λ_{GML} is smaller than $\lambda^* = O(1/n^{q/(pq+1)})$, and an easy calculation shows that $\lim_{n\to\infty}(ET(\lambda_{\mathrm{GML}})/ET(\lambda^*)) \uparrow \infty$. On the other hand, it is argued in Wahba (1985e) that, if f is a sample function from a stochastic process with $Ef(s)f(t) = R(s,t)$, then the minimizers of both $V(\lambda)$ and $M(\lambda)$ estimate σ^2/nb. Thus, it is inadvisable to use the maximum likelihood estimate of λ since it is not robust against deviations from the stochastic model.

4.9 Limits of GCV.

The theory justifying the use of GCV is an asymptotic one. Good results cannot be expected for very small sample sizes when there is not enough information in the data to separate signal from noise. To take an extreme example, imagine, say, $n = 5$ data points (y_i, x_i) with x_i on the line. For arbitrary scattered values of the y_i's, given no further information, a curve interpolating the points, or the least squares straight line regression to the points, could be equally reasonable, and $V(\lambda)$ may well have minima at both zero and infinity. However, if even the order of magnitude of σ^2 is known in an example like this, then one could likely decide between these two extremes. My own Monte Carlo studies with smooth "truth" and independent and identically distributed Gaussian noise have resulted in generally reliable estimates of λ for n upwards of 25 or 30. It is to be noted that even for larger n, say $n = 50$, in extreme Monte Carlo replications there may be a handful of unwarranted extreme estimates ($\hat{\lambda} = 0$ or $\hat{\lambda} = \infty$), say a few percent, while the remaining estimates are all reasonable and more or less clustered together. This effect has been noted in Wahba (1983) and Section 6.3. Generally, if only σ^2 is known to within an order of magnitude, the occasional extreme case can be readily identified. As n gets larger, this effect becomes weaker, although it still defies ordinary statistical intuition. Even with "nice" examples with $n = 200$, there may be an occasional (2 or 3 out of 1,000, say) outliers in an otherwise "pleasant" population of sample $\hat{\lambda}$'s. One imagines that the theoretical distribution of $\hat{\lambda}$ can have (small) mass points at $\lambda = 0$ and $\lambda = \infty$ for moderate n.

My experience with GCV is that it is fairly robust against nonhomogeneity of variances and non-Gaussian errors (see, e.g., Villalobos and Wahba (1987)), and appears to work well when the ϵ_i's are due to quantization (see, e.g., Shahrary and Anderson (1989)). Andrews (1988) has recently provided some favorable theoretical results for unequal variances. However, the method is quite likely to give unsatisfactory results if the errors are highly correlated. It has given poor results when used to smooth a sample cumulative distribution F_n, for example, where $F_n(x_i) - F(x_i)$ and $F_n(x_j) - F(x_j)$ are correlated (Nychka, 1983) whereas differencing the data (see, e.g., Nychka et al. (1984)) so that the ϵ_i's are nearly independent has given good results. In a recent thesis, Altman (1987) discusses GCV in the presence of correlated errors. Of course if the noise is highly correlated, it becomes harder to distinguish it from "signal" by any nonparametric method that does not "know" anything about the nature of the correlation.

Trouble can arise with GCV if one has "exact" data (i.e., $\sigma^2 = 0$) *and* some of the $\lambda_{\nu n}$ appearing in (4.6.3) are insufficiently distinguishable from machine zero even though (in theory) they are strictly positive. In this case the theoretically "right" λ is zero, but in practice the numerical calculations with $\lambda = 0$ or λ near machine 0 can cause numerical instabilities and an unsatisfactory solution. Behavior of the $\lambda_{\nu n}$ in some well-known problems is discussed later.

CHAPTER 5

"Confidence Intervals"

5.1 Bayesian "confidence intervals."

Continuing with the Bayesian model

$$\begin{aligned} F(t) &= \sum_{\nu=1}^{M} \theta_\nu \phi_\nu(t) + b^{1/2} X(t), \ t \in \mathcal{T}, \\ Y_i &= L_i F + \epsilon_i \end{aligned}$$

as in (1.5.8), we know that

$$\lim_{a \to \infty} E(F(t) | Y_i = y_i, i = 1, \ldots, n) = f_\lambda(t)$$

with $\lambda = \sigma^2/nb$. The covariance of $f_\lambda(s)$ and $f_\lambda(t)$, call it $c_\lambda(s,t)$, can be obtained by standard multivariate techniques. A formula is given in Wahba (1983), which we do not reproduce here. (This formula also involves b.)

By the arguments in Section 1.5,

$$E(L_0 F | y_1, \ldots, y_n) = L_0 f_\lambda,$$

and it is not hard to show that the covariance of $L_0 f_\lambda$ and $L_{00} f_\lambda$ is $L_{0(s)} L_{00(t)} c_\lambda(s,t)$.

An important special case that will be used to construct "confidence intervals" is: The covariance matrix of $(L_1 f_\lambda, \ldots, L_n f_\lambda)$ is

$$\text{cov}\ (L_1 f_\lambda, \ldots, L_n f_\lambda) = \sigma^2 A(\lambda). \tag{5.1.1}$$

One way to derive (5.1.1) is to consider the Bayes model of (1.5.8) before letting $a \to \infty$. Then the joint covariance matrix of $(L_1 F, \ldots, L_n F, Y_1, \ldots, Y_n)$ is

$$\begin{pmatrix} aTT' + b\Sigma & aTT' + b\Sigma \\ & \\ aTT' + b\Sigma & aTT' + b\Sigma \ \ +\sigma^2 I \end{pmatrix}.$$

Then we have

$$\begin{pmatrix} L_1 f_\lambda \\ \vdots \\ L_n f_\lambda \end{pmatrix} = \lim_{a \to \infty} A^a(\lambda) y$$

where $A^a(\lambda) = (aTT' + b\Sigma)(aTT' + b\Sigma + \sigma^2 I)^{-1}$, with $\lambda = \sigma^2/nb$. It can be verified using the limit formulae (1.5.11) and (1.5.12) that $\lim_{a\to\infty} A^a(\lambda) = A(\lambda)$. Then

$$\begin{pmatrix} L_1 f_\lambda - L_1 F \\ \vdots \\ L_n f_\lambda - L_n F \end{pmatrix} = -\lim_{a\to\infty} (I - A^a(\lambda)) \begin{pmatrix} L_1 F \\ \vdots \\ L_n F \end{pmatrix} + A^a(\lambda)\epsilon,$$

with covariance

$$\lim_{a\to\infty} [(I - A^a(\lambda))(aTT' + b\Sigma)(I - A^a(\lambda)) + \sigma^2 A^a(\lambda) \cdot A^a(\lambda)]. \tag{5.1.2}$$

The collection of terms in (5.1.2) shows that the quantity in brackets in (5.1.2) is equal to $\sigma^2 A^a(\lambda)$, giving the result.

Considering the case $L_i F = F(t_i)$, we have that (5.1.1) suggests using as a confidence interval

$$f_{\hat\lambda}(t_i) \pm z_{\alpha/2}\sqrt{\hat\sigma^2 a_{ii}(\hat\lambda)},$$

where $\hat\lambda$ and $\hat\sigma^2$ are appropriate estimates of λ and σ^2 and $z_{\alpha/2}$ is the $\alpha/2$ point of the normal distribution.

The estimate

$$\hat\sigma^2 = \frac{\text{RSS}(\hat\lambda)}{\text{Trace}\,(I - A(\hat\lambda))} \tag{5.1.3}$$

where $\text{RSS}(\hat\lambda)$ is the residual sum of squares, was used in the example below. Although to the author's knowledge theoretical properties of this estimate have not been published, good numerical results in simulation studies have been found by several authors (see, e.g., O'Sullivan and Wong (1988), Nychka (1986a,b, 1988), Hall and Titterington (1987)). The argument is that trace $A(\hat\lambda)$ should be considered the degrees of freedom (d.f.) for signal, by analogy with the regression case, and trace $(I - A(\hat\lambda))$ is the d.f. for noise. On a hunch, it was decided to study these "confidence intervals" numerically with smooth functions and the GCV estimates $\hat\lambda$ of λ.

Figure 5.1 gives a test function from Wahba (1983). Data were generated according to the model

$$y_{ij} = f\left(\frac{2i+1}{2N}, \frac{2j+1}{2N}\right) + \epsilon_{ij}, i, j = 1, \ldots, N$$

with $N = 13$, giving $n = N^2 = 169$ data points. The peak height of f was approximately 1.2 and σ was taken as .03. $f_{\hat\lambda}$ was the thin-plate spline of Section 2.4 with $d = 2, m = 2$. Figure 5.2 gives four selected cross sections for four fixed values of $x_1, x_1 = (2i+1)/N$, for $i = 7, 9, 11, 13$. In each cross section is plotted $f((2i+1)/N, x_2)$, $0 \le x_2 \le 1$ (solid line), $f_{\hat\lambda}((2i+1)/N, x_2)$, $0 \le x_2 \le 1$, where $f_{\hat\lambda}$ is the thin plate smoothing line (dashed line), the data y_{ij}, $j = 1, \ldots, 13$, for i fixed, and confidence bars, which extend between

$$f_{n,\hat\lambda}((2i+1)/N, x_2(j)) \pm 1.96\hat\sigma(\hat\lambda)\sqrt{a_{ij,ij}(\hat\lambda)}.$$

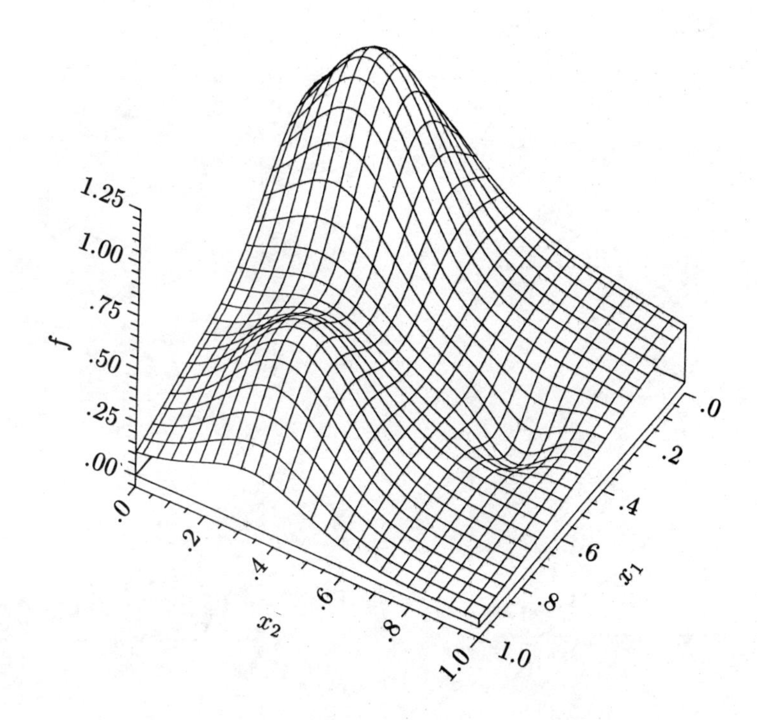

FIG. 5.1. *Test function for confidence intervals.*

Of the 169 confidence intervals, 162 or 95.85 percent covered the true value of $f(x_1(i), x_2(j))$.

We take pains to note that these "confidence intervals" must be interpreted "across the function," as opposed to pointwise. If this experiment were repeated with the same f and new ϵ_{ij}'s then it would be likely that about 95 percent of the confidence intervals would cover the corresponding true values, but it may be that the value at the same (x_1, x_2) is covered each time. This effect is more pronounced if the true curve or surface has small regions of particularly rapid change. In an attempt to understand why these Bayesian confidence intervals have the frequentist properties that they apparently do, it was shown that

$$ET(\lambda^*) = \alpha \frac{\sigma^2}{n} \sum_{i=1}^{n} a_{ii}(\lambda^*)(1 + o(1)), \tag{5.1.4}$$

where λ^* is the minimizer of $ET(\lambda^*)$ and, in the case of the univariate polynomial spline of degree $2m-1$ with equally spaced data, $\alpha \in [(1+1/4m)(1-1/2m), 1]$, that is, $(\sigma^2/n)\,\mathrm{Tr}\,A(\lambda^*)$ is actually quite close to $ET(\lambda^*) = b^2(\lambda^*) + \sigma^2\mu_2(\lambda^*)$. Nychka (1986a,b, 1988) and Hall and Titterington (1987) later showed that the lower bound obtained, and Nychka gave a nice argument rigorizing the interpretation of intervals as confidence intervals "across the function" by

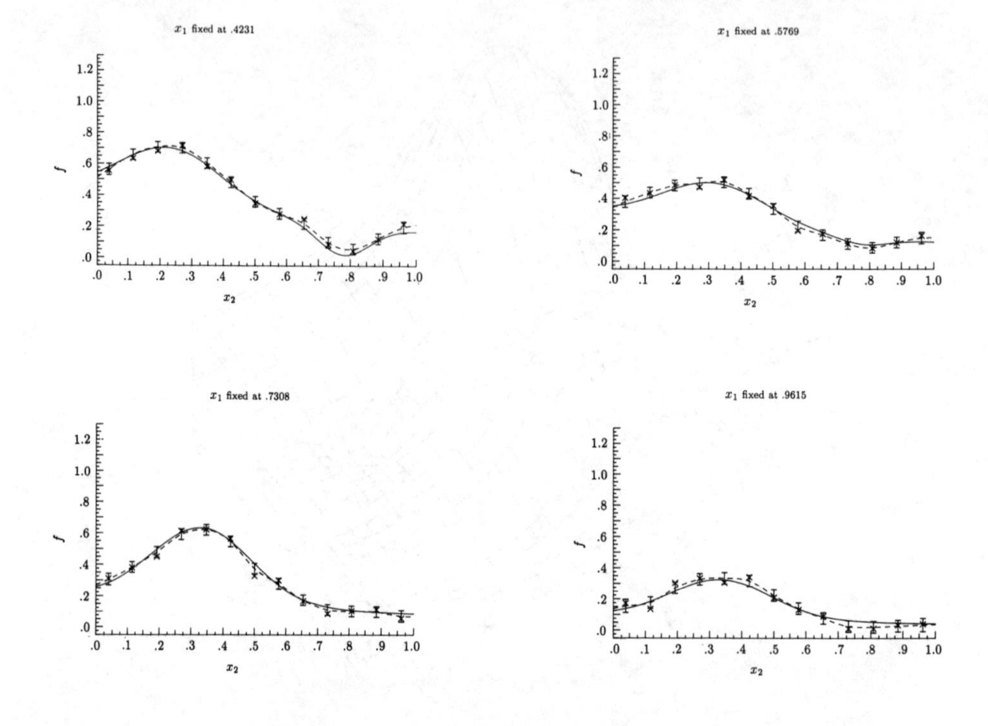

FIG. 5.2. *Cross sections of $f, f_{\hat{\lambda}}$, and "confidence intervals."*

working with the "average coverage probability" (ACP), defined by

$$\frac{1}{n}\sum_{i=1}^{n} P\{f(t_i) \in C_{.95}(t_i)\}$$

where $C_{.95}(t_i)$ is the ith confidence interval. These and similar confidence intervals have also been discussed by Silverman (1984) and O'Sullivan (1986a).

Nychka (see also Shiau (1985)) argued that a confidence interval based on the distribution $\mathcal{N}(0, \mu_i^2 + \sigma_i^2)$ should not be, in a practical sense, too far from correct when the true distribution is $N(\mu_i, \sigma_i^2)$, provided that μ_i^2 is not large compared to σ_i^2. Here let $\mu_i = Ef_{\hat{\lambda}}(t_i) - f(t_i)$ and $\sigma_i^2 = E(f_{\hat{\lambda}}(t_i) - Ef_{\hat{\lambda}}(t_i))^2$. We are, "on the average," replacing $\mathcal{N}(\mu_i, \sigma_i^2)$ with $\mathcal{N}(0, \mu_i^2 + \sigma_i^2)$, since $1/n\sum_{i=1}^{n}(\mu_i^2 + \sigma_i^2) = b^2(\hat{\lambda}) + \sigma^2\mu_2(\hat{\lambda})$ $\simeq b^2(\lambda^*) + \sigma^2\mu_2(\lambda^*) = \alpha\sigma^2/n\sum_{i=1}^{n} a_{ii}(\lambda^*)(1 + o(1))$ by (5.1.4). The minimization of $T(\lambda)$ with respect to λ entails that the square bias $b^2(\lambda^*)$ be of the same order as the variance $\sigma^2\mu_2(\lambda^*)$. In examples it tends to be of moderate size with respect to the variance.

Considering the case of univariate spline smoothing in W_m,we remark that if f is going to be in $\mathcal{H}_{R^p}$ of (4.5.4) for some $p > 1$, then f must satisfy some boundary conditions. For example, let $p = 1 + k/m$ for some $k \le m$; then $f \in \mathcal{H}_{R^{1+k/m}}$ if $f^{(k+m)} \epsilon \mathcal{L}_2$ and $f^{(\nu)}(0) = f^{(\nu)}(1) = 0$ for $\nu = m, m+1, \ldots, m+k-1$. Thus f can be "very smooth" in the interior of $[0,1]$ in the sense that $f^{(k+m)} \epsilon \mathcal{L}_2$, but if f does not satisfy the additional boundary conditions, then the higher-order convergence rates will not hold (see Rice and Rosenblatt (1983)).

In the case of confidence intervals, if f is "very smooth" in the interior, but fails to satisfy the higher-order boundary conditions, this would tend to cause the 5 percent of coverage failures for 95 percent confidence intervals to repeatedly fall near the boundary. This is similar to the way that the failed confidence intervals tend to repeat over a break in the first derivative of the true f if it occurs in the interior of $[0,1]$. (See Wahba (1983) for examples of this.) Nychka (1988) has proposed procedures for excluding the boundaries.

5.2 Estimate-based bootstrapping.

Another approach to confidence interval estimation may be called "estimate-based bootstrapping" (see also Efron (1982), Efron and Tibshirani (1986)). It goes as follows. From the data obtain $f_{\hat{\lambda}}$ and $\hat{\sigma}^2(\hat{\lambda})$, then, pretending that $f_{\hat{\lambda}}$ is the "true" f, generate data

$$\tilde{y}_i = f_{\hat{\lambda}}(t_i) + \tilde{\epsilon}_i,$$

where $\tilde{\epsilon}_i \sim \mathcal{N}(0, \hat{\sigma}^2(\hat{\lambda}))$, from a random number generator. (Here we are supposing that $L_i f = f(t_i)$.) Then find $\tilde{f}_{\tilde{\lambda}}$, based on the data $\tilde{y}$. Upon repeating this calculation l times (with l different $\tilde{\epsilon}$), one has a distribution of l values of $\tilde{f}_{\tilde{\lambda}}(t_i)$ at each t, and the $\alpha/2$ lth and $(1-\alpha/2)$ lth values can be used for a "confidence interval" (see, for example O'Sullivan (1988a)). The properties of these "confidence intervals" are not known. Plausible results have been obtained in simulation experiments, however. It is possible that the results will be too "rosy," since $f_{\hat{\lambda}}$ can be expected to display less "fine structure" than f. It would be a mistake to take the raw residuals $\hat{\epsilon}_i(\hat{\lambda}) = y_i - f_{\hat{\lambda}}(t_i)$, $i = 1, \ldots, n$ and generate data by

$$\tilde{y}_i = f_{\hat{\lambda}}(t_i) + \tilde{\epsilon}_i$$

where $\tilde{\epsilon}_i$ is drawn from the population $\{\hat{\epsilon}_i(\hat{\lambda}), \ldots, \hat{\epsilon}_n(\hat{\lambda})\}$, since $1/n \sum \hat{\epsilon}_i^2(\hat{\lambda}) = \text{RSS}(\hat{\lambda}) \simeq \sigma^2/\text{Tr}\ (I - A(\hat{\lambda}))$, the $\tilde{\epsilon}_i$ should be corrected for d.f. noise first if this approach were to be used.

Other important diagnostic tools are discussed in Eubank (1984, 1985).

CHAPTER 6

Partial Spline Models

6.1 Estimation.

As before, let $\mathcal{H} = \mathcal{H}_0 \oplus \mathcal{H}_1$, where $\mathcal{H}_0$ is M-dimensional, and let $L_1, \ldots, L_n$ be n bounded linear functionals on $\mathcal{H}$. Let $\psi_1, \ldots, \psi_q$ be q functions such that the $n \times q$ matrix S with irth entry

$$S_{ir} = L_i \psi_r$$

is well defined and finite. Letting the matrix $T_{n \times M}$ have $i\nu$th entry $L_i \phi_\nu$ as before, where $\phi_1, \ldots, \phi_M$ span $\mathcal{H}_0$, we will need to suppose that the $n \times (M+q)$ matrix

$$X = (S : T) \tag{6.1.1}$$

is of full column rank (otherwise there will be identifiability problems). The original abstract spline model was

$$y_i = L_i f + \epsilon_i, \quad i = 1, \ldots, n$$

$$f \epsilon \mathcal{H}_0 \oplus \mathcal{H}_1.$$

Find $f_\lambda \in \mathcal{H}$ to minimize

$$\frac{1}{n} \sum_{i=1}^{n} (y_i - L_i f)^2 + \lambda \|P_1 f\|_{\mathcal{H}}^2.$$

The partial spline model is

$$y_i = \sum_{r=1}^{q} \beta_r L_i \psi_r + L_i f + \epsilon_i, \quad i = 1, \ldots, n \tag{6.1.2}$$

where

$$f \in \mathcal{H} = \mathcal{H}_0 \oplus \mathcal{H}_1,$$

as before. Now we find $\beta = (\beta_1, \ldots, \beta_q)'$ and $f \epsilon \mathcal{H}$ to minimize

$$\frac{1}{n} \sum_{i=1}^{n} \left(y_i - \sum_{r=1}^{q} \beta_r L_i \psi_r - L_i f \right)^2 + \lambda \|P_1 f\|_{\mathcal{H}}^2. \tag{6.1.3}$$

I originally believed I was the first to think up these wonderful models around 1983 while enjoying the hospitality of the Berkeley Mathematical Sciences Research Center. Their generation was an attempt to extend the applicability of thin-plate splines to semiparametric modeling of functions of several variables with limited data, and the result appears in Wahba (1984b, 1984c). The work was also motivated by the ideas in Huber (1985) on projection pursuit concerning "interesting directions." I soon found that the idea of partial splines, which has a wealth of applications, had occurred to a number of other authors in one form or another—Ansley and Wecker (1981), Anderson and Senthilselvan (1982), Shiller (1984), Laurent and Utreras (1986), Eubank (1986), Engle, Granger, Rice and Weiss (1986), to mention a few. (The work of Laurent and Utreras and Engle et al. appears earlier in unpublished manuscripts in 1980 and 1982, respectively.)

The application of Engle et al. is quite interesting. They had electricity sales y_i billed each month i for four cities, over a period of years. They also had price ψ_1, income ψ_2, and average daily temperatures x, for each month, by city. The idea was to model electricity demand h as the sum of a smooth function f of monthly temperature x, and linear functions of ψ_1 and ψ_2, along with 11 monthly dummy variables $\psi_3, \ldots, \psi_{13}$, that is, the model was

$$h(x, \psi_1, \ldots, \psi_{13}) = \sum_{\nu=1}^{13} \beta_r \psi_r + f(x)$$

where f is "smooth."

Engle et al. did not observe the daily electricity demand directly, but only certain weighted averages $L_i h$ of it resulting from the fact that the total monthly sales billed reflected the staggered monthly billing cycles. Thus, their model was

$$y_i = \sum_{r=1}^{13} \beta_r \; L_i \psi_r + L_i f + \epsilon_i, \quad i = 1, 2, \ldots, n.$$

An additional twist of their model was the assumption that, rather than being independent, the ϵ_i's followed a first-order autoregressive scheme

$$\epsilon_i = \rho \epsilon_{i-1} + \delta_i,$$

where the δ_i's are independently and identically distributed for some ρ. This assumption appeared reasonable in the light of the staggered data collection scheme. For the right ρ the quasi-differences

$$\tilde{y}_i = y_i - \rho y_{i-1}$$

result in a new model with independent errors

$$\tilde{y}_i = \sum_{r=1}^{13} \beta_r \tilde{L}_i \psi_r + \tilde{L}_i f + \delta_i, \quad i = 1, \ldots, n,$$

where $\tilde{L}_i = L_i - \rho L_{i-1}$. They fit the model (6.1.2) with $\|P_1 f\|^2 = \int (f''(x))^2 dx$, using gridpoint discretization, whereby the function f is approximated by a vector of its values on a (fine) grid. $\tilde{L}_i f$ is a linear combination of the values of f, $\int (f''(x))^2 dx$ is replaced by a sum of squares of second divided differences, and so forth. The influence matrix $A(\lambda)$ can be found and the GCV estimate $\lambda = \lambda_\rho$ found. Little detail was given on their selection of ρ but Engle et al. noted that all of their estimates for ρ appeared to be quite similar. Altman (1987) has reiterated that care must be taken when the errors are correlated and has studied in depth some procedures appropriate in that case, including the selection of ρ.

Returning to the abstract partial spline model, from the geometric point of view of Kimeldorf and Wahba (1971), we have not done anything new, except adjoin span $\{\psi_r\}_{r=1}^q$ to $\mathcal{H}$, giving a new Hilbert space $\tilde{\mathcal{H}}$

$$\tilde{\mathcal{H}} = \mathcal{H}_{00} \oplus \mathcal{H}_0 \oplus \mathcal{H}_1,$$

where $\mathcal{H}_{00} \equiv \text{span}\{\psi_r\}$. Then $\tilde{\mathcal{H}}_0 = \mathcal{H}_{00} \oplus \mathcal{H}_0$ is the (new) null space of the penalty functional. By the same argument as in Kimeldorf and Wahba, one shows that

$$h = \sum_{r=1}^{q} \beta_r \psi_r + \sum_{\nu=1}^{M} d_\nu \phi_\nu + \sum_{i=1}^{n} c_i \xi_i, \tag{6.1.4}$$

and the problem becomes: Find β, c, d to minimize

$$\frac{1}{n}\|y - S\beta - Td - \Sigma c\|^2 + \lambda c' \Sigma c.$$

Letting $\alpha = \begin{pmatrix} \beta \\ d \end{pmatrix}$, we get

$$\frac{1}{n}\|y - X\alpha - \Sigma c\|^2 + \lambda c' \Sigma c,$$

$$(\Sigma + n\lambda I)\, c + X\alpha = y,$$

$$X'c = 0,$$

and the GCV estimate of λ can be obtained as before.

We did not say anything concerning properties of the functions $\psi_1, \ldots, \psi_q$, other than the fact that the $L_i\psi_r$ must be well defined and the columns of $X = (S : T)$ must be linearly independent with $q + M \le n$. It does not otherwise matter whether or not the ψ_r are, say, in $\mathcal{H}$ as the following way of looking at the problem from a geometric point of view will show.

Let $\tilde{\mathcal{H}}$ be a Hilbert space with elements

$$h = (h_0, f_0, f_1)$$

where $h_0 \epsilon \mathcal{H}_{00} \equiv$ span $\{\psi_r\}, f_0 \epsilon \mathcal{H}_0$ and $f_1 \epsilon \mathcal{H}_1$, $\mathcal{H}_0$ and $\mathcal{H}_1$ being as before. We define the projection operators P_{00}, P_0, and P_1 in $\tilde{\mathcal{H}}$ as

$$\begin{aligned} P_{00}h &= (h_0, 0, 0), \\ P_0 h &= (0, f_0, 0), \\ P_1 h &= (0, 0, f_1), \end{aligned}$$

and the squared norm

$$\begin{aligned}\|h\|_{\tilde{\mathcal{H}}}^2 &= \|h_0\|_{\mathcal{H}_{00}}^2 + \|f_0\|_{\mathcal{H}_0}^2 + \|f_1\|_{\mathcal{H}_1}^2 \\ &= \|P_{00}h\|_{\tilde{\mathcal{H}}}^2 + \|P_0 h\|_{\tilde{\mathcal{H}}}^2 + \|P_1 h\|_{\tilde{\mathcal{H}}}^2.\end{aligned}$$

By convention $L_i h = L_i h_0 + L_i f_0 + L_i f_1$. Now consider the following three problems.

PROBLEM 1. Find $h_\lambda^{(1)}$, the minimizer in $\tilde{\mathcal{H}}$ of

$$\frac{1}{n}\sum_{i=1}^n (y_i - L_i h)^2 + \lambda \|P_1 h\|_{\tilde{\mathcal{H}}}^2.$$

PROBLEM 2. Find $\hat{\beta}^{(2)}$ and $h_\lambda^{(2)}$, the minimizer in $(E^q, \tilde{\mathcal{H}})$ of

$$\frac{1}{n}\sum_{i=1}^n \left(y_i - \sum_{r=1}^q \beta_r L_i \psi_r - L_i h \right)^2 + \lambda \|(P_{00} + P_1) h\|_{\tilde{\mathcal{H}}}^2.$$

PROBLEM 3. Find $\hat{\beta}^{(3)}$, $d^{(3)}$, and $h_\lambda^{(3)}$, the minimizer in $(E^q, E^M, \tilde{\mathcal{H}})$ of

$$\frac{1}{n}\sum_{i=1}^n \left(y_i - \sum_{r=1}^q \beta_r L_i \psi_r - \sum_{\nu=1}^M d_\nu L_i \phi_\nu - L_i h \right)^2 + \lambda \|h\|_{\tilde{\mathcal{H}}}^2.$$

It is not hard to convince oneself that, if

$$h_\lambda^{(1)} = \left(\sum_{r=1}^q \hat{\beta}_r \psi_r, \sum_{\nu=1}^M \hat{d}_\nu \phi_\nu, \hat{f}_1 \right)$$

then

$$h_\lambda^{(2)} = \left(0, \sum_{\nu=1}^M \hat{d}_\nu \phi_\nu, \hat{f}_1 \right), \ \hat{\beta}^{(2)} = \hat{\beta}$$

and

$$h_\lambda^{(3)} = \left(0, 0, \hat{f}_1 \right), \ \hat{\beta}^{(3)} = \hat{\beta}, \ \hat{d}^{(3)} = \hat{d},$$

where

$$\hat{\beta} = (\hat{\beta}_1, \ldots, \hat{\beta}_q)', \ \hat{d} = (\hat{d}_1, \ldots, \hat{d}_M)'.$$

What this says is that explicitly representing an element of a particular subspace in the sum of squares term effectively puts it in the null space of the penalty functional, whether or not it is there already.

Another important application of partial spline models is to model a function of one or several variables as a function that is smooth except for a discontinuity in a low-order derivative at a specific location. Here, let $f \in W_m$ and let

$$h(x) = \sum_{r=1}^q \beta_r \psi_r(x) + f(x)$$

where $\psi_r(x) = (x - x_r)_+^{q_r}$. Here h will have jumps in its derivatives at x_r,

$$\lim_{x \downarrow x_r} h^{(q_r)}(x) - \lim_{x \uparrow x_r} h^{(q_r)}(x) = \beta_r \cdot q_r!$$

This problem was discussed by Ansley and Wecker (1981), and Laurent and Utreras (1986). Shiau (1985) considered various classes of jump functions in several variables, and Shiau, Wahba, and Johnson (1986) considered a particular type of jump function in two dimensions that is useful in modeling two-dimensional atmosphere temperature (as a function of latitude and height, say) where it is desired to model the sharp minimum that typically occurs at the tropopause.

Figure 6.1 from Shiau, Wahba, and Johnson (1986), gives a plot of atmospheric temperature $h(z,l)$ as a function of height z and latitude l. In keeping with meteorological convention, this figure is tipped on its side. The model was

$$h(z,l) = \beta\psi(z,l) + f(z,l)$$

where f is a thin plate spline and

$$\psi(z,l) = |z - z^*(l)|.$$

$z^*(l)$ is shown in Figure 6.2 and $\psi(z,l)$ is shown in Figure 6.3.

6.2 Convergence of partial spline estimates.

We will only give details for $q = 1$. The results below can be extended to the general case of $q << n - M$. We want to obtain a simple representation for $\hat{\beta} = \hat{\beta}_\lambda$, to examine the squared bias and variance. The calculations below follow Shiau (1985), Heckman (1986), and Shiau and Wahba (1988). The model is

$$y_i = \beta L_i\psi + L_i f + \epsilon_i, \ i = 1, \ldots, n \tag{6.2.1}$$

and the partial spline estimator of β and $f \in \mathcal{H}$ is the minimizer of

$$\frac{1}{n}\sum_{i=1}^{n}(y_i - \beta L_i\psi - L_i f)^2 + \lambda\|P_1 f\|^2. \tag{6.2.2}$$

Let $s = (L_1\psi, \ldots, L_n\psi)'$ and $A_0(\lambda)$ be the influence matrix for the problem (6.2.2) if β is identically zero. It is easy to see that, for any fixed β,

$$\begin{pmatrix} L_1 f_\lambda \\ \vdots \\ L_n f_\lambda \end{pmatrix} = A_0(\lambda)(y - s\beta),$$

by minimizing (6.2.2) with $(y - s\beta)$ as "data."

As in Section 4.5, letting $P_1 f_\lambda = \Sigma c_i \xi_i$, we have $\|P_1 f_\lambda\|^2 = c'\Sigma c$, and, since $n\lambda c = (I - A_0(\lambda))(y - s\beta)$, after some algebra, we obtain that $n\lambda c'\Sigma c = (y - s\beta)' A_0(\lambda)(I - A_0(\lambda))(y - s\beta)$ and (6.2.2) is equal to

$$\frac{1}{n}(y - s\beta)'(I - A_0(\lambda))(y - s\beta).$$

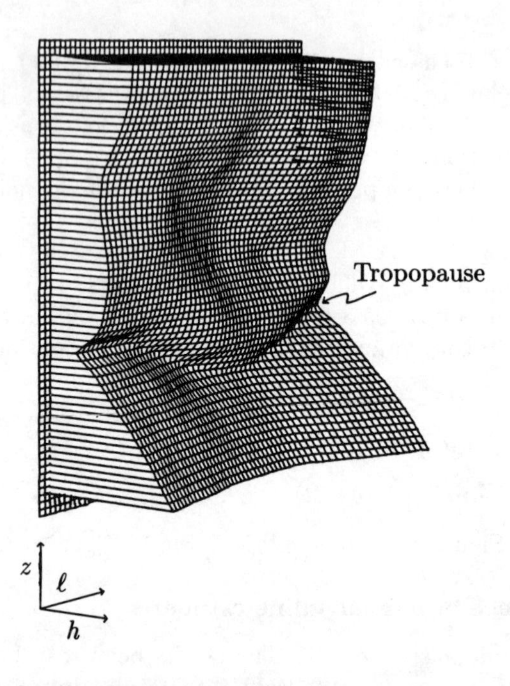

FIG. 6.1. *Estimated temperature $h(z, \ell)$ as a function of height z and latitude ℓ.*

Minimizing over β (and relying on the assumption that $A_0 s \neq s$, which follows from the assumption that s and the columns of T are linearly independent, recall that the the columns of T are the eigenvectors of A_0 with unit eigenvalue), we have

$$\hat{\beta}_\lambda = (s'(I - A_0)s)^{-1}s'(I - A_0)y. \tag{6.2.3}$$

Letting $g = (L_1 f, \ldots, L_n f)'$, we have

$$\hat{\beta}_\lambda - \beta = \frac{s'(I - A_0)(g + \epsilon)}{s'(I - A_0)s} \tag{6.2.4}$$

so that

$$\begin{aligned} \text{bias}\,(\hat{\beta}_\lambda) &= \frac{s'(I - A_0)g}{s'(I - A_0)s} \\ \text{var}\,(\hat{\beta}_\lambda) &= \sigma^2 \frac{s'(I - A_0)^2 s}{[s'(I - A_0)s]^2}. \end{aligned} \tag{6.2.5}$$

Let A_0 be as in (1.3.23) and (4.4.2), that is,

$$I - A_0(\lambda) \;=\; n\lambda Q_2(Q_2'(\Sigma + n\lambda I)Q_2)^{-1}Q_2',$$

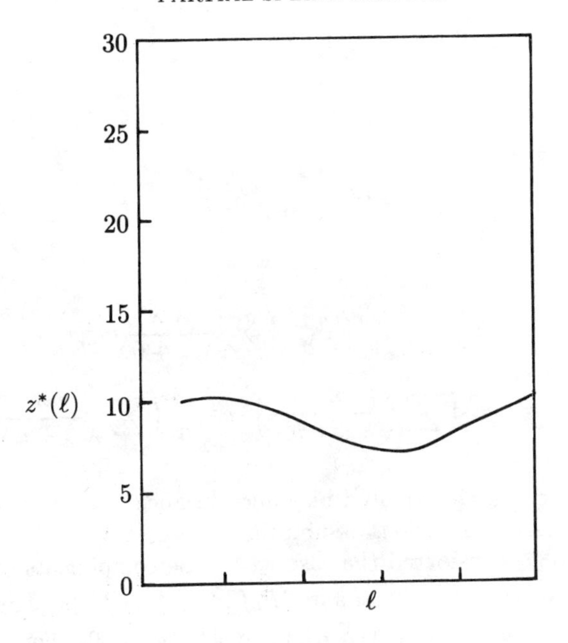

FIG. 6.2. *The tropopause,* $z^*(\ell)$.

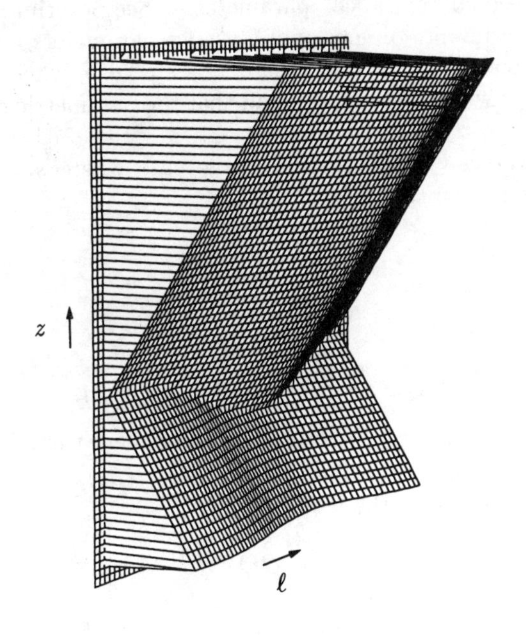

FIG. 6.3. *The tropopause break function* $\psi(z, \ell)$.

$$\begin{aligned} Q_2'\Sigma Q_2 &= UDU', \\ \Gamma &= Q_2U. \end{aligned}$$

Let

$$h = \Gamma' g, \quad u = \Gamma' s.$$

Then we have

$$\begin{aligned} (\text{bias})\,(\hat{\beta}_\lambda) &= \sum \frac{(n\lambda)u_{\nu n}h_{\nu n}}{\lambda_{\nu n} + n\lambda} \Big/ \left(\sum \frac{n\lambda}{\lambda_{\nu n} + n\lambda} u_{\nu n}^2 \right) \qquad (6.2.6) \\ \text{var}\,(\hat{\beta}_\lambda) &= \sigma^2 \left(\sum \left(\frac{n\lambda}{\lambda_{\nu n} + n\lambda} \right)^2 u_{\nu n}^2 \Big/ \left(\sum \frac{n\lambda}{\lambda_{\nu n} + n\lambda} u_{\nu n}^2 \right)^2 \right). \end{aligned}$$

Asymptotic theory for the squared bias and variance of $\hat{\beta}_\lambda$ has been developed by several authors under various assumptions.

Heckman (1986) considered the case where the components of s (and hence the $u_{\nu n}$) behaved as white noise, and $\|P_1 f\|^2 = \int_0^1 (f^{(m)}(u))^2\,du$. She proved that if $n\lambda^{1/2m} \to \infty$ and either $\lambda \to 0$, or $\|P_1 f\|^2 = 0$, then $\sqrt{n}(\hat{\beta}_\lambda - \beta)$ is asymptotically normal with mean zero and finite variance. See also Chen (1988). In this case we note that a parametric rate for MSE$(\hat{\beta}_\lambda)$ is obtained in the case of an infinite-dimensional "nuisance" parameter g. See Severini and Wong (1987) for a discussion on infinite-dimensional nuisance parameters.

Rice (1986) considered the case where $s_i = s(t_i) + \delta_i$ where $s(\cdot)$ is "smooth" and δ_i behaves like white noise, and found that the parametric rate $1/\sqrt{n}$ cannot hold.

Shiau and Wahba (1988) considered the case where $s_i = s(t_i)$ with s a "smooth" function, such that

$$u_{\nu n} \simeq \sqrt{n}\nu^{-p}, \qquad (6.2.7)$$

and

$$h_{\nu n} \simeq \sqrt{n}\nu^{-q},$$

$$\lambda_{\nu n} \simeq n\nu^{-2m}, \quad \nu = 1, \ldots, n - M.$$

We also assumed that $q > p > \frac{1}{2}$ so that s is "rougher" than f, and $s \in \mathcal{L}_2$.

To analyze the behavior of bias^2 $(\hat{\beta}_\lambda)$ and var $(\hat{\beta}_\lambda)$, one substitutes (6.2.7) into (6.2.6) and sums, using the lemma

$$\sum_{\nu=1}^{n} \left(\frac{\lambda \nu^{r-\theta}}{(1 + \lambda\nu^r)} \right)^k = \begin{cases} O(\lambda^{(k\theta-1)/r}) & \text{if } \frac{k\theta-1}{k} < r \\ O(\lambda^k) & \text{if } \frac{k\theta-1}{k} > r, \end{cases} \qquad (6.2.8)$$

good for $r > 0, \theta > 0$ and $k \geq 1$. Some of the results are given in Table 6.1. The third column of Table 6.1, λ_{opt}, is the rate of decay for λ that minimizes MSE $(\hat{\beta}_\lambda) = \text{bias}^2(\hat{\beta}_\lambda) + \text{var}(\hat{\beta}_\lambda)$, and the fourth column is the MSE for $\lambda = \lambda_{\text{opt}}$.

TABLE 6.1
Bias, variance, λ_{opt}*, and* MSE (λ_{opt}) *for* $\hat{\beta}_\lambda$.

	$\text{bias}^2(\hat{\beta}_\lambda)$	$\text{var}(\hat{\beta}_\lambda)$	$\lambda_{\text{opt}}(\hat{\beta}_\lambda)$	$\text{MSE}_{\lambda_{\text{opt}}}(\hat{\beta}_\lambda)$
$2m > p+q-1$	$\lambda^{\frac{2(q-p)}{2m}}$	$(n\lambda^{\frac{2p-1}{2m}})^{-1}$	$n^{-\frac{2m}{2q-1}}$	$n^{\frac{-2(q-p)}{2q-1}}$
$p+q-1 > 2m > 2p-1$	$\lambda^{\frac{2(2m-2p+1)}{2m}}$	$(n\lambda^{\frac{2p-1}{2m}})^{-1}$	$n^{-\frac{2m}{4m-2p+1}}$	$n^{\frac{-2(2m-2p+1)}{4m-2p+1}}$
$2p-1 > 2m$	$O(1)$			

We now compare the rates of convergence of λ^* and λ_{opt}, where λ^* is the optimal rate for minimizing the predictive mean-square error $T(\lambda)$ in smoothing splines. We have

$$
\begin{aligned}
ET(\lambda) &= \frac{1}{n}E\|g + s\beta - \hat{g}_\lambda - s\hat{\beta}_\lambda\|^2 \\
&= \frac{1}{n}E\|(g+s\beta) - A_0(\lambda)(g + s\beta + \epsilon - s\hat{\beta}_\lambda) - s\hat{\beta}_\lambda\|^2 \\
&= \frac{1}{n}E\|(I - A_0(\lambda))(g - s(\hat{\beta}_\lambda - \beta)) - A_0(\lambda)\epsilon\|^2 \\
&= \frac{1}{n}E\|(I - A_0(\lambda))(g - s(s'(I - A_0(\lambda))s)^{-1}s'(I - A_0(\lambda))g) \\
&\qquad -(I - A_0(\lambda))s(s'(I - A_0(\lambda))s)^{-1}(s'(I - A_0(\lambda))\epsilon) - A_0(\lambda)\epsilon\|^2 \\
&= \frac{1}{n}\|(I - A_0(\lambda))(g - s\cdot(\text{bias}(\hat{\beta}_\lambda)))\|^2 \text{ (squared bias term)} \\
&+ \frac{\sigma^2}{n}\{\text{tr}A_0^2(\lambda) + 2s'(I - A_0(\lambda))A_0(\lambda)(I - A_0(\lambda))s[s'(I - A_0(\lambda))s]^{-1} \\
&+ s'(I - A_0(\lambda))^2 s \ \text{var}(\hat{\beta}_\lambda)/\sigma^2\} \text{ (variance term).}
\end{aligned}
$$

We just consider the case $2m - 2q + 1 > 0$. Then under the assumption we have made on the $h_{\nu n}$ and $\lambda_{\nu n}$, we have that the two main terms in the squared bias term are:

$$\frac{1}{n}\|(I - A_0(\lambda))g\|^2 = O(\lambda^{(2q-1)/2m}),$$

$$
\begin{aligned}
\frac{1}{n}\|(I - A_0(\lambda))s\|^2 \,\text{bias}^2(\hat{\beta}_\lambda) &= O(\lambda^{(2p-1)/2m}) \cdot O(\lambda^{2(q-p)/2m}) \\
&= O(\lambda^{(2q-1)/2m})
\end{aligned}
$$

and it can be shown that the variance term is dominated by

$$\frac{\sigma^2}{n} \,\text{tr}\, A_0^2(\lambda) = O(n^{-1}\lambda^{-1/2m}).$$

If there is no cancellation in the squared bias term, then we get

$$\text{Squared bias} + \text{variance} = O(\lambda^{(2q-1)/2m}) + O(n^{-1}\lambda^{-1/2m})$$

and

$$\lambda^* = O(n^{-2m/2q}),$$

whereas from Table 6.1, we have

$$\lambda_{\text{opt}}(\hat{\beta}_\lambda) = O(n^{-2m/(2q-1)}) \text{ when } 2m > p + q - 1.$$

So that in this case, $\lambda_{\text{opt}}(\hat{\beta}_\lambda)$ goes to zero at a faster rate than λ^*.

Speckman (1988) and Denby (1986) have proposed an estimator $\tilde{\beta}_\lambda$ that can have a faster convergence rate than $\hat{\beta}_\lambda$ of (6.2.3). It was motivated as follows.

Let $\tilde{y} = (I - A_0(\lambda))y$ be the residuals after removing the "smooth" part $A_0(\lambda)y$, and let $\tilde{s} = (I - A_0(\lambda))s$. Then by regressing $\tilde{y}$ on $\tilde{s}$, the Denby–Speckman estimate $\tilde{\beta}_\lambda$ of β is obtained:

$$\tilde{\beta}_\lambda = (\tilde{s}'\tilde{s})^{-1}\tilde{s}'\tilde{y} = [s'(I - A_0(\lambda))^2 s]^{-1} s'(I - A_0(\lambda))^2 y.$$

Formulas analogous to (6.2.6) for $\tilde{\beta}_\lambda$ are obvious. Table 6.2 gives the square bias, variance λ_{opt}, and MSE (λ_{opt}) for $\tilde{\beta}_\lambda$, from Shiau and Wahba (1988).

TABLE 6.2
Bias, variance λ_{opt}, and MSE_{opt} for $\tilde{\beta}_\lambda$.

	$\text{bias}^2(\tilde{\beta}_\lambda)$	$\text{var}(\tilde{\beta}_\lambda)$	$\lambda_{\text{opt}}(\tilde{\beta}_\lambda)$	$\text{MSE}_{\lambda\ \text{opt}}(\tilde{\beta}_\lambda)$
$2m > \frac{p+q-1}{2}$	$\lambda^{\frac{2(q-p)}{2m}}$	$(n\lambda^{\frac{2p-1}{2m}})^{-1}$	$n^{-\frac{2m}{2q-1}}$	$n^{-\frac{2(q-p)}{2q-1}}$
$\frac{p+q-1}{2} > 2m > \frac{2p-1}{2}$	$\lambda^{\frac{2(4m-2p+1)}{2m}}$	$(n\lambda^{\frac{2p-1}{2m}})^{-1}$	$n^{-\frac{2m}{8m-2p+1}}$	$n^{-\frac{2(4m-2p+1)}{8m-2p+1}}$
$\frac{2p-1}{2} > 2m$	$O(1)$			

It can be seen by comparison of Tables 6.1 and 6.2 that there are combinations of p, q, and m for which $\text{MSE}_{\lambda_{\text{opt}}}(\tilde{\beta}_\lambda)$ is of smaller order than $\text{MSE}_{\lambda_{\text{opt}}}(\hat{\beta}_\lambda)$, and combinations for which they are the same order (see Shiau and Wahba (1988) for more details).

6.3 Testing.

Returning to our Bayes model of

$$y_i = \sum_{\nu=1}^{M} \theta_\nu L_i \phi_\nu + b^{1/2} L_i X + \epsilon_i, \; i = 1, \ldots, n, \tag{6.3.1}$$

we wish to test the null hypothesis

$$b = 0, \ y_i = \sum_{\nu=1}^{M} \theta_\nu L_i \phi_\nu + \epsilon_i, \ i = 1, \dots, n \tag{6.3.2}$$

versus the alternative

$$b \neq 0. \tag{6.3.3}$$

Letting $T_{n \times M} = \{L_i \phi_\nu\}$, and $\Sigma = \{L_{i(s)} L_{j(t)} Q(s,t)\}$, where $EX(s)X(t) = Q(s,t)$, we have

$$y \sim \mathcal{N}(T\theta, b\Sigma + \sigma^2 I)$$

under the "fixed effects" model, and

$$y \sim \mathcal{N}(0, \xi TT' + b\Sigma + \sigma^2 I)$$

under the "random effects" model.

The most interesting special case of this is

$$y_i = f(x_i) + \epsilon_i, \quad i = 1, \dots, n$$

with the null hypothesis f a low-degree polynomial, versus the alternative, f "smooth." Yanagimoto and Yanagimoto (1987) and Barry and Hartigan (1988) considered maximum likelihood tests for this case. Cox and Koh (1986) considered the case $f \in W_m$ and obtained the locally most powerful (LMP) invariant test. Cox, Koh, Wahba, and Yandell (1988) considered the LMP invariant test in the general case, and obtained a relation between the LMP invariant and the GCV test (to be described). In preparation for the CBMS conference, I obtained a similar relation for the GML test (to be described) and did a small Monte Carlo study to compare the LMP invariant, GML, and GCV tests.

Let $T = (Q_1 : Q_2) \begin{pmatrix} R \\ 0 \end{pmatrix}$ as in (1.3.18). As before

$$T'Q_2 = 0.$$

Let

$$w_1 = Q_1' y, \qquad w_2 = Q_2' y;$$

then

$$w_2 \sim \mathcal{N}(0, bQ_2' \Sigma Q_2 + \sigma^2 I)$$

for either the fixed or random effects model. Letting $UDU' = Q_2' \Sigma Q_2$ and

$$z = U' w_2$$

we obtain

$$z \sim \mathcal{N}(0, bD + \sigma^2 I), \tag{6.3.4}$$

where the diagonal entries of D are $\lambda_{\nu n}, \nu = 1, \ldots, n-M$, that is

$$z_\nu \sim \mathcal{N}(0, b\lambda_{\nu n} + \sigma^2), \; \nu = 1, \ldots, n-M.$$

Cox and Koh (1986) showed that the LMP test (at $b = 0$), invariant under translation by columns of T, rejects when

$$t_{\mathrm{LMP}} = \sum_{\nu=1}^{n-M} \lambda_{\nu n} z_\nu^2 \tag{6.3.5}$$

is too large.

A test based on the GCV estimate for λ may be obtained by recalling that $\lambda = \infty$ corresponds to $f_\lambda \epsilon \mathcal{H}_0$. In the notation of this section, we have

$$\begin{aligned} n^{-1}V(\lambda) &= \frac{\|(I - A(\lambda))y\|^2}{(\mathrm{tr}(I - A(\lambda)))^2} \\ &= \sum_{\nu=1}^{n-M} \left(\frac{n\lambda}{n\lambda + \lambda_{\nu n}}\right)^2 z_\nu^2 \Big/ \left(\sum_{\nu=1}^{n-M} \frac{n\lambda}{n\lambda + \lambda_{\nu n}}\right)^2 . \end{aligned} \tag{6.3.6}$$

We have the following theorem.

THEOREM 6.3.1. *$V(\lambda)$ has a (possibly local) minimum at $\lambda = \infty$, whenever*

$$t_{\mathrm{LMP}} \equiv \sum_{\nu=1}^{n-M} \lambda_{\nu n} z_\nu^2 \leq \bar{\lambda} \sum_{\nu=1}^{n-M} z_\nu^2, \tag{6.3.7}$$

where $\bar{\lambda} = 1/(n-M)\sum_{\nu=1}^{n-M} \lambda_{\nu n}$.

Proof. Let $\gamma = 1/n\lambda$, and define $\tilde{V}(\gamma)$ as $V(\lambda)$ with $1/n\lambda$ replaced by γ, that is,

$$n^{-1}\tilde{V}(\gamma) = \sum_{\nu=1}^{n-M} \left(\frac{1}{1 + \gamma\lambda_{\nu n}}\right)^2 z_\nu^2 \Big/ \left(\sum_{\nu=1}^{n-M} \frac{1}{1 + \gamma\lambda_{\nu n}}\right)^2 . \tag{6.3.8}$$

Differentiating $\tilde{V}(\gamma)$ with respect to γ, one obtains that

$$\tilde{V}'(\gamma)\Big|_{\gamma=0} \geq 0 \text{ iff } \sum_{\nu=1}^{n-M} \lambda_{\nu n} z_\nu^2 \leq \bar{\lambda} \sum_{\nu=1}^{n-M} z_\nu^2.$$

We note that $\sum_{\nu=1}^{n-M} z_\nu^2$ is the residual sum of squares after regression of y onto the columns of T, that is, the residual sum of squares after fitting the null model.

An approximate LMP invariant test when σ^2 is unknown is to use

$$t_{\mathrm{LMP\ approx}} = \sum_{\nu=1}^{n-M} \lambda_{\nu n} z_\nu^2 \Big/ \sum_{\nu=1}^{n-M} z_\nu^2.$$

The GCV test is

$$t_{\mathrm{GCV}} = \text{const.}\ \frac{\tilde{V}(\hat{\gamma})}{\tilde{V}(0)} = \frac{\sum\left(z_\nu^2/(1+\hat{\gamma}\lambda_{\nu n})^2\right)}{\left(\sum\left(1/(1+\hat{\gamma}\lambda_{\nu n})\right)^2\right)} \times \frac{1}{\sum z_\nu^2}$$

where $\hat{\gamma} = 1/n\hat{\lambda}$.

The (invariant) likelihood ratio test statistic t_{GML} for $\lambda = \sigma^2/nb = \infty$ is

$$t_{\mathrm{GML}} = \text{ const.}\ \frac{M(\tilde{\lambda})}{M(\infty)},$$

where $\tilde{\lambda}$ minimizes $M(\lambda)$ of (4.8.4). Upon letting $\tilde{\gamma} = 1/n\tilde{\lambda}$, we have

$$t_{\mathrm{GML}} = \frac{\sum\left(z_\nu^2/(1+\tilde{\gamma}\lambda_{\nu n})\right)}{\prod(1+\tilde{\gamma}\lambda_{\nu n})^{-1/(n-M)}} \times \frac{1}{\sum z_\nu^2}.$$

It can also be shown that $M(\lambda)$ has a (possibly local) minimum at $\lambda = \infty$, whenever (6.3.7) holds.

An experiment was designed to examine the relative power of $t_{\mathrm{LMP}}, t_{\mathrm{GCV}}$, and t_{GML} by simulation. It was to be expected that t_{LMP} would have the greatest power for nearby alternatives but would not be so good for "far out" alternatives, and it was tentatively conjectured that t_{GML} would be better than GCV for "random" alternatives (to be defined) but GCV would be better for "smooth" alternatives. We considered the one-dimensional cubic smoothing spline with $n = 100$ data points at $x_i = i/n$; there are 98 eigenvalues $\lambda_{\nu n}$, which are plotted in Figure 6.4. This corresponds to the null hypothesis that f is linear. We also considered a two-dimensional thin-plate spline with $m = 2$ on a 12×12 regular grid with $(x_1, x_2) = (i/12, j/12)$, $i = j = 1, \ldots, 12$, thus $n = 144$, $n - M = 141$. The null hypothesis corresponds to $f(x_1, x_2)$, a plane. The 141 eigenvalues are also plotted in Figure 6.4. The eigenvalues $\lambda_{\nu n}$ for these splines decay at the rate $\nu^{-2m/d}$ where $m = 2$ here and d is the dimension of x; here $d = 1$ and $d = 2$. The decay rate of the eigenvalues is readily evident.

The random alternative was

$$z_\nu \sim \mathcal{N}(0, b\lambda_{\nu n} + \sigma^2)$$

where the size of b controlled the distance of the alternative from the null hypothesis, and, without loss of generality we set $\sigma^2 = 1$. The "smooth" fixed function corresponded to

$$z_\nu \sim \mathcal{N}(\sqrt{b}h_{\nu n}, \sigma^2)$$

with $\sum(h_{\nu n}^2/\lambda_{\nu n}) < \infty$. Here the $h_{\nu n}$ are related to $f(x)$ by

$$\begin{pmatrix} h_{1n} \\ \vdots \\ h_{n-M,n} \end{pmatrix} = U'Q_2' \begin{pmatrix} f(x(1)) \\ \vdots \\ f(x(n)) \end{pmatrix}.$$

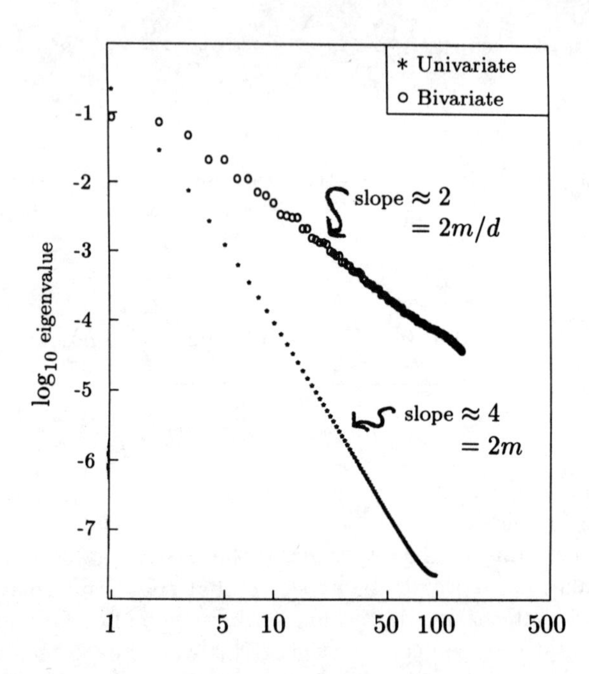

FIG. 6.4. *Univariate* ($n = 98$) *and bivariate* ($n = 141$) *eigenvalues.*

Figure 6.5 gives a plot of the univariate alternative $f(x)$ at $x(i)$, $i = 1, \ldots, n$ with $b = 1$; actually f was chosen so that the first four $h_{\nu n}$'s were 1,1,-1, and 1 and the rest 0. Figure 6.6 gives a plot of the bivariate smooth alternative $f(x_1, x_2)$. This function was chosen so that the first three $h_{\nu n}$'s were 1 and the rest 0.

The distributions of $t_{\mathrm{LMP}}, t_{\mathrm{GCV}}$, and t_{GML} for the univariate example under the null hypothesis were estimated by drawing 1,000 replicates of $n - M = 98$ $\mathcal{N}(0,1)$ z_ν's, and computing 1,000 values of each statistic. Global search in increments of $\log \gamma$ was used for the minimizations of the GCV and GML functions. Care must be taken that the search increment is sufficiently fine and over a sufficiently wide range.

Figures 6.7 and 6.8 give histograms of $-\log t_{\mathrm{GCV}}$ and $-\log t_{\mathrm{GML}}$ under the null hypothesis. If V or M is minimized for $\gamma = 1/\lambda = \infty$, then $-\log t_{\mathrm{GCV}}$ or $-\log t_{\mathrm{GML}}$ is zero. We note that these were 581 samples of $-\log t_{\mathrm{GCV}} = 0$ and 628 samples of $-\log t_{\mathrm{GML}} = 0$. Defining $S/N = (b \sum \lambda_{\nu n}/n\sigma^2)^{1/2}$ and $(b \sum h_{\nu n}^2/n\sigma^2)^{1/2}$ for the random and smooth alternatives, respectively, 1,000 replicates of $t_{\mathrm{LMP}}, t_{\mathrm{GCV}}$, and t_{GML} for a series of values of S/N were generated. The same 98 $\times$ 1,000 random numbers were used for the different values of S/N, so the data in Figures 6.7–6.14 are not independent. The histograms

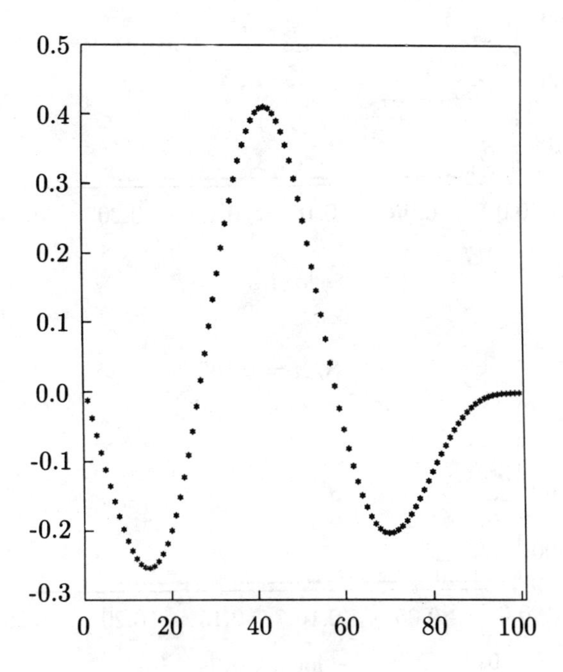

FIG. 6.5. *The univariate alternative.*

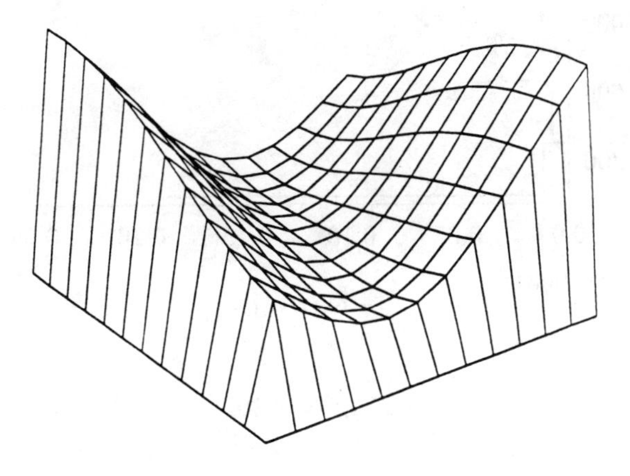

FIG. 6.6. *The bivariate alternative.*

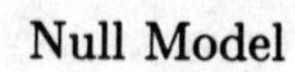

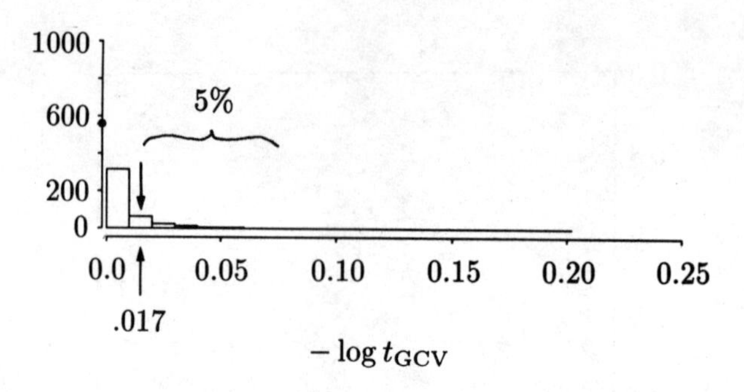

S/N = 0.102

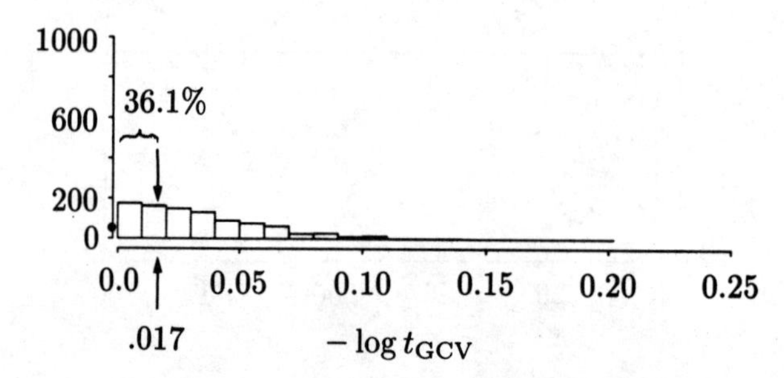

FIG. 6.7. *Histograms for* $-\log t_{\mathrm{GCV}}$, *univariate deterministic example.*

Null Model

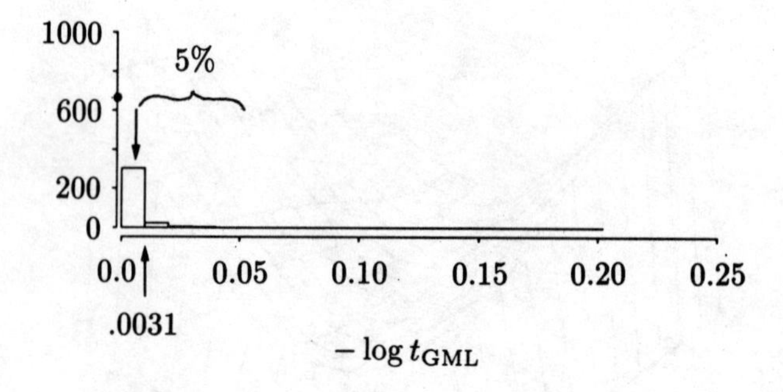

S/N = 0.102

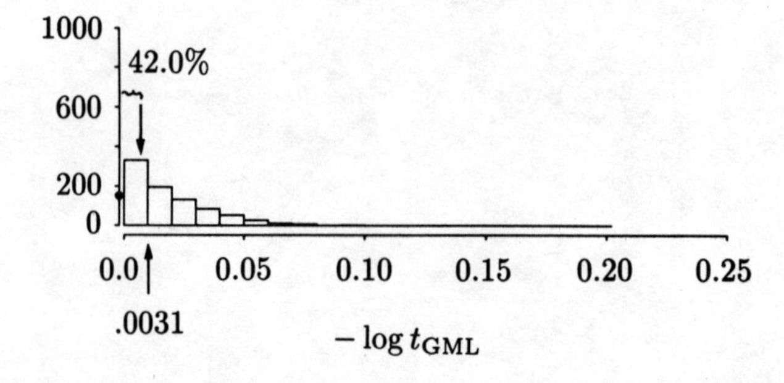

FIG. 6.8. *Histograms for* $-\log t_{\mathrm{GML}}$, *univariate deterministic example.*

for $-\log t_{\mathrm{GCV}}$ and $-\log t_{\mathrm{GML}}$ for the univariate smooth function case for $S/N = .102$ appear in Figures 6.7 and 6.8. It can be seen that a nonnegligible mass point appears at zero. Using the 95 percent points of the simulated null distributions as cutoff points, the probability of accepting the null hypothesis for various values of S/N was estimated by counting how many of the 1,000 simulated values of t fell below the cutoff. These estimated type-two errors are plotted for the three statistics for the univariate smooth alternative in Figure 6.9, for the univariate random alternative in Figure 6.10, and for the bivariate smooth alternative in Figure 6.11. It appears that t_{GCV} is slightly better in the deterministic example and t_{GML} in the random example, but we do not believe these experiments are definitive. The sampling error is fairly large (and its magnitude is not evident in the plots because the same random numbers were used for different S/N), and in other experiments the reverse was found. The results can also be surprisingly sensitive to the search procedure used.

Figures 6.12–6.14 show the histograms for the three test statistics for six values of S/N. We remark that in practice Monte Carlo estimation of distributions such as these can be important.

The distributions of the test statistics $-\log t_{\mathrm{GCV}}$ and $-\log t_{\mathrm{GML}}$ have a mass point at zero, which shrinks as S/N becomes larger. This mass point is not plotted separately in the histograms of Figures 6.12 and 6.13, but is displayed in Figures 6.7 and 6.8. The continuous part of some of the distributions appears to have a bimodal density, although to see the exact behavior near zero probably would require a more delicate search procedure than we have used. Simple asymptotic approximations to the distributions are, in our opinion, not necessarily trustworthy for practical use due to the complex structure of these distributions.

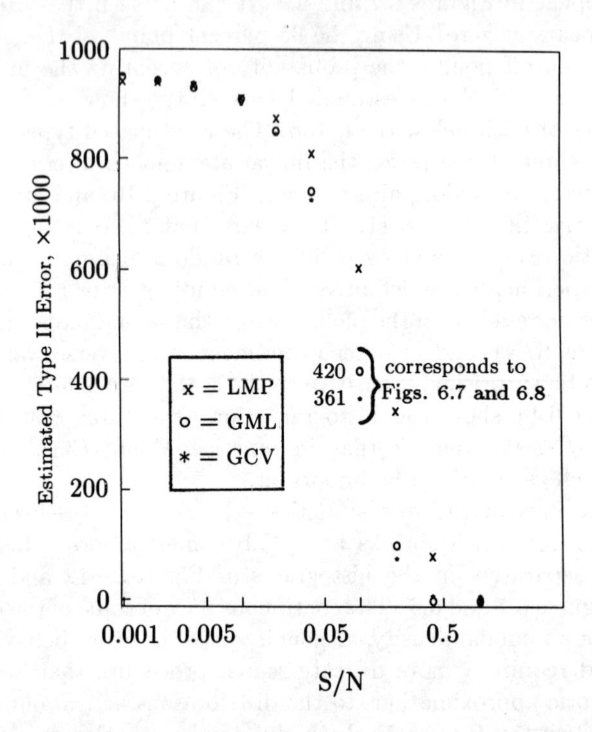

FIG. 6.9. *Univariate example, smooth deterministic alternative.*

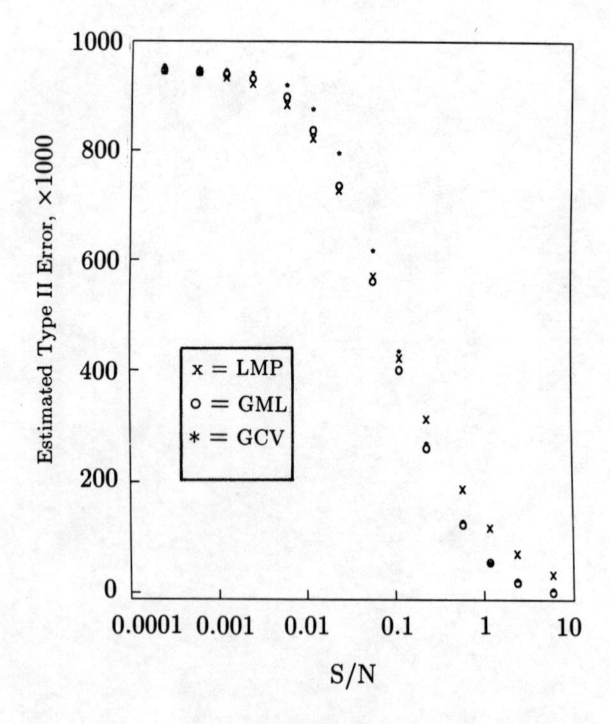

FIG. 6.10. *Univariate example, random alternative.*

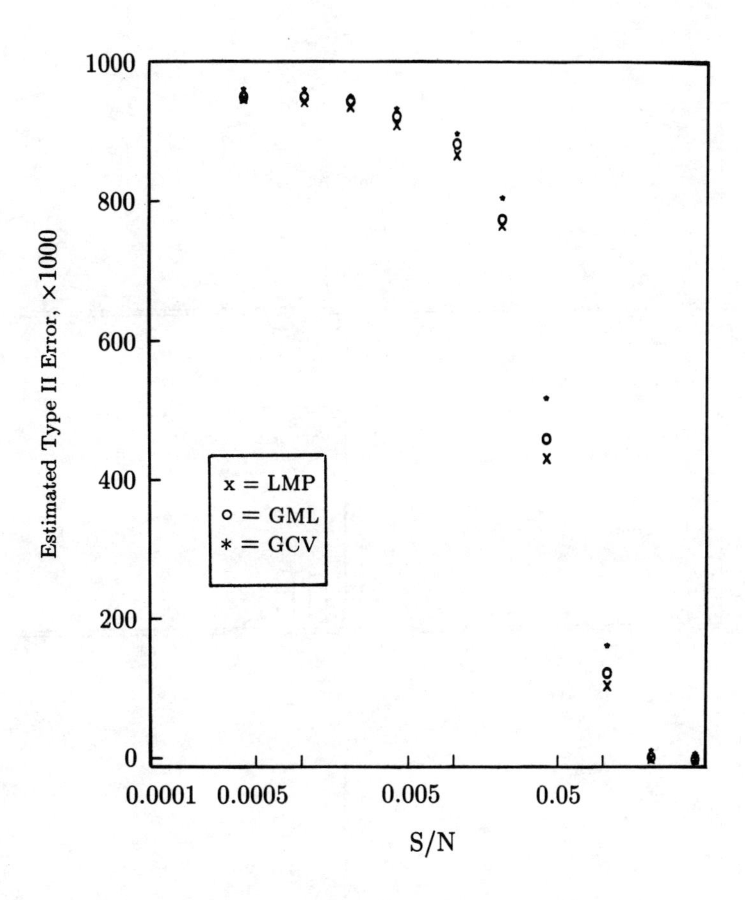

FIG. 6.11. *Bivariate example, deterministic alternative.*

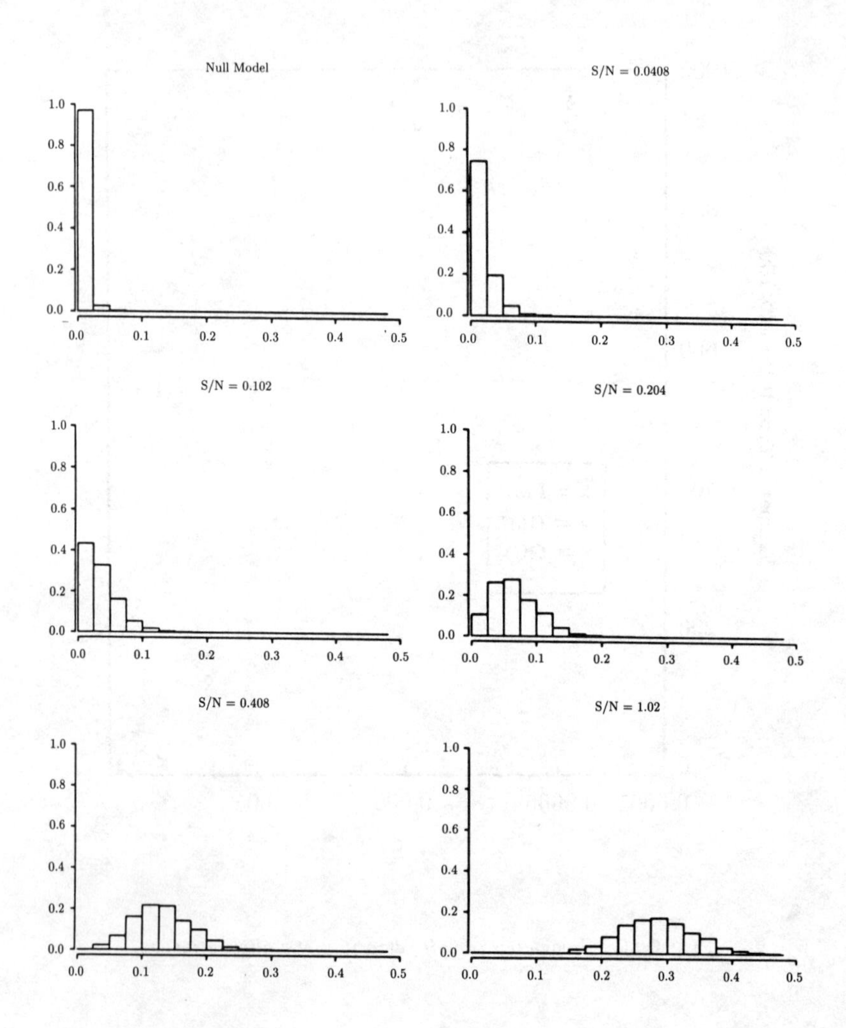

FIG. 6.12. *Histograms for* $-\log t_{\text{GCV}}$, *univariate deterministic example.*

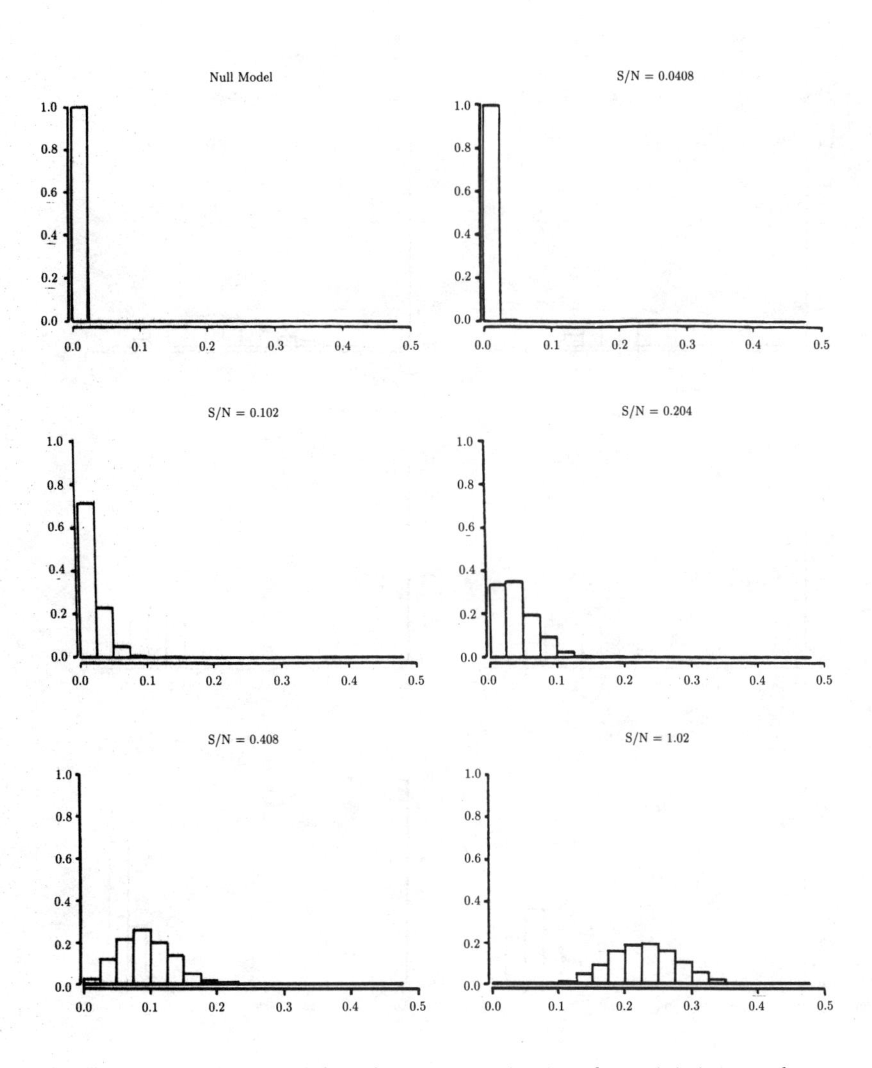

FIG. 6.13. *Histograms for* $-\log t_{\text{GML}}$, *univariate deterministic example.*

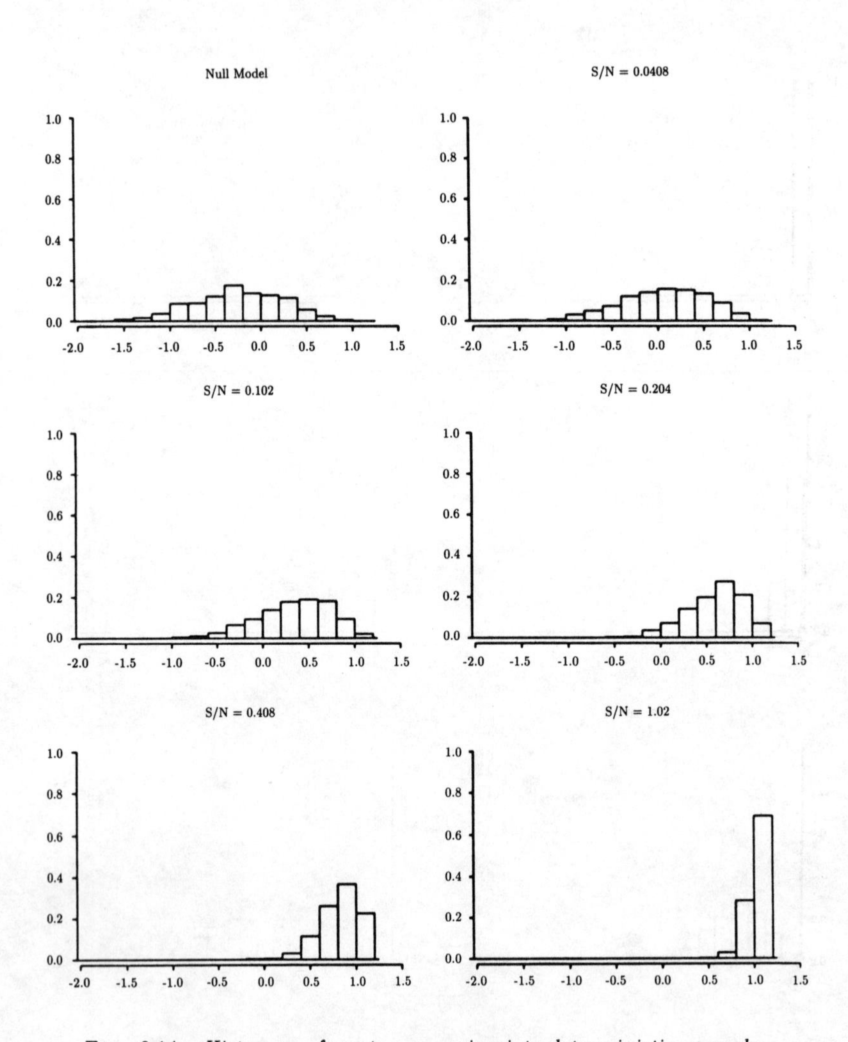

FIG. 6.14. *Histograms for* $-t_{\text{LMP}}$*, univariate deterministic example.*

CHAPTER 7

Finite-Dimensional Approximating Subspaces

7.1 Quadrature formulae, computing with basis functions.

Suppose we wish to compute the minimizer in $\mathcal{H}_R$ of

$$\frac{1}{n}\sum_{i=1}^{n}(y_i - L_i f)^2 + \lambda \|P_1 f\|^2 \tag{7.1.1}$$

where either n is very large, and/or, we do not have a closed form expression for

$$\xi_i(t) = L_{i(u)} R^1(t,u).$$

If, for example,

$$L_i f = \int_\Omega K(t_i, u) f(u)\, du,$$

then

$$\xi_i(t) = \int_\Omega K(t_i,u) R^1(t,u)\, du, \tag{7.1.2}$$

and, if a closed form expression is not available for $\xi_i(t)$, it would appear that a quadrature or other approximation to (7.1.2) would be necessary.

A quadrature formula in the context of $\mathcal{H}_R$ can be obtained as follows. Let $s_1, \ldots, s_N$ be N (distinct) points in $\mathcal{T}$ such that the $N \times M$ matrix with $l\nu$th entry $\phi_\nu(s_l)$ is of rank M, and, for any $f \in \mathcal{H}_R$, let f_0 be that element in $\mathcal{H}_R$ that minimizes $\|P_1 f_0\|^2$ subject to $f_0(s_l) = f(s_l)$, $l = 1, \ldots, N$. Then, if η is the representer of integration in $\mathcal{H}_R$

$$< \eta, f >= \int f(s)\, ds, \tag{7.1.3}$$

we approximate $< \eta, f >$ by $< \eta, f_0 >$, which is a linear combination of the values of f at $s_1, \ldots, s_N$ (i.e., a quadrature formula). Certain classical quadrature formulae can be obtained this way (see Schoenberg (1968)). Now define $\hat{\eta}$ by the relationship $< \hat{\eta}, f >=$ $< \eta, f_0 >$, all $f \in \mathcal{H}_R$. Since $< \hat{\eta}, f >$ depends on f only through $f(s_1), \ldots, f(s_N)$, it follows that $\hat{\eta}$ is in span $\{R_{s_1}, \ldots, R_{s_N}\}$, where R_{s_l} is the representer of evaluation at s_l. Thus $< \hat{\eta}, f - f_0 >= 0$, for any $f \in \mathcal{H}_R$. It can also be shown that $< \eta - \hat{\eta}, f >= 0$ for any $f \in \mathcal{H}_0$, that is, the quadrature approximation is exact for $f \in \mathcal{H}_0$. This is equivalent to $\eta - \hat{\eta} \perp \mathcal{H}_0$. These facts result in the so-called hypercircle inequality (see Golomb and Weinberger (1959), Wahba (1969)), which goes as follows. Since $< \hat{\eta}, f >=< \eta, f_0 >$, we have

$$< \eta - \hat{\eta}, f >=< \eta, f - f_0 >=< \eta - \hat{\eta}, f - f_0 >=< \eta - \hat{\eta}, P_1(f - f_0) >$$

$$=< P_1(\eta - \hat{\eta}),\ P_1(f - f_0) >$$

so that

$$| < \eta, f > - < \hat{\eta}, f > | \leq \|P_1(\eta - \hat{\eta})\| \cdot \|P_1(f - f_0)\|.$$

High-order convergence rates for $\|P_1(f - f_0)\|$ are available in the context of univariate splines (see Schultz (1973a), Schumaker (1981)), and thin-plate splines (Duchon (1978)). The famous "optimal quadrature" problem can be formulated as the problem of choosing $s_1, \ldots, s_N$ to minimize $\|P_1(\eta - \hat{\eta})\|$ or $\sup_{f \in \mathcal{E}} \|P_1(f - f_0)\|$ for some class $\mathcal{E}$ (see Schoenberg (1968) and Section 12.2).

This kind of quadrature was discussed in Nychka et al. (1984) in the context of numerically minimizing (7.1.1) and it was noted that essentially the same accuracy can be obtained in a computationally much simpler way by minimizing (7.1.1) in a certain subspace $\mathcal{H}_N$ spanned by N suitably chosen basis functions. Given basis functions $B_1, \ldots, B_N$, one sets

$$f = \sum_{k=1}^{N} c_k B_k,$$

substitutes this expression into (7.1.1), and solves for the coefficients $c_1, \ldots, c_N$. We next discuss the choice of basis functions.

In Wahba (1980b) it was proposed, in the context of $L_i f = f(t_i)$ with very large n, that a good subspace $\mathcal{H}_N$ of basis functions can be chosen as follows. Choose $s_1, \ldots, s_N$ points distributed "nicely" over $\mathcal{T}$, and let the basis functions $B_1, \ldots, B_N$ be chosen as follows. Let $B_1, \ldots, B_M$ be $\phi_1, \ldots, \phi_M$. Let $u_l = (u_{1l}, \ldots, u_{Nl})'$, $l = 1, 2, \ldots, N - M$ be $N - M$ linearly independent vectors with the property

$$\sum_{k=1}^{N} u_{kl} \phi_\nu(s_k) = 0,\ l = 1, \ldots, N - M,\ \nu = 1, \ldots, M \qquad (7.1.4)$$

and let

$$B_{M+l} = \sum_{k=1}^{N} u_{kl} P_1 R_{s_k},\ l = 1, \ldots, N - M. \qquad (7.1.5)$$

We have that for any $f \in \mathcal{H}_R$, f_0, that element in $\mathcal{H}_R$, which minimizes $\|P_1 f_0\|^2$ subject to $f_0(s_l) = f(s_l)$, $l = 1, \ldots, N$, is in $\mathcal{H}_N$. It so happens in $W_m[0,1]$ (but *not* in general), that there exist coefficients u_{kl} in (7.1.5) so that the B_{M+l}, $l = 1, 2, \ldots, N - M$ have compact support. This special case is important, so we will describe it in some detail. The coefficients are those that characterize ordinary divided differences. We show this next. Here, $M = m$ and we let $s_1 < s_2 < \ldots < s_N$. Using the reproducing kernel for $W_m[0,1]$ given in Section 1.2 we have, for any fixed s,

$$P_1 R_s(t) = \int_0^1 \frac{(s-u)_+^{m-1}(t-u)_+^{m-1}}{[(m-1)!]^2}\, du = \xi_s(t), \text{ say.}$$

Recall that, for fixed s, ξ_s, considered as a function of t, satisfies

$$\xi_s \in \pi^{2m-1}, \ t \in [0,s],$$

$$\xi_s \in \pi^{m-1}, \ t \in [s,1],$$

ξ_s has $2m-2$ continuous derivatives, and $\xi_s(t)$ is symmetric in s and t. Let $[s_l, \ldots, s_{l+2m}]\xi_s$ denote the $2m$th divided difference of ξ_s with respect to s, for example, a first divided difference is $[s_1, s_2]\xi_s = (\xi_{s_2} - \xi_{s_1})/(s_2 - s_1)$. Let

$$B_{m+l} = [s_l, \ldots s_{l+2m}]\xi_s, \ l = 1, \ldots, N-2m.$$

Then B_{m+l} (considered as a function of t) is a linear combination of $\xi_{s_l}, \xi_{s_{l+1}}, \ldots, \xi_{s_{l+2m}}$. B_{m+l} is hence a piecewise polynomial of degree at most $2m-1$ with knots at $s_l, \ldots, s_{l+2m}$, and possessing $2m-2$ continuous derivatives. We next show that $B_{m+l}(t) = 0$ for $t \notin [s_l, s_{l+2m}]$. For any fixed $t \le s_l \le s$ we may write

$$\xi_s(t) = \sum_{\nu=0}^{m-1} s^\nu f_\nu(t), \quad s \ge s_l \ge t,$$

for some f_ν's, since $\xi_s(t)$ is a polynomial of degree $m-1$ in s for $s \ge t$. Similarly, for $t \ge s_{l+2m} \ge s$ we may write

$$\xi_s(t) = \sum_{\nu=0}^{2m-1} s^\nu \tilde{f}_\nu(t)$$

for some $\tilde{f}_\nu$. Since $[s_l, \ldots, s_{l+2m}]s^r = 0$ for any $r = 1, 2, \ldots, 2m-1$, it follows that

$$[s_l, \ldots, s_{l+2m}]\xi_s(t) \equiv B_{m+l}(t) = 0, \quad t \notin [s_l, s_{l+2m}]$$

This gives $N-2m$ basis functions with compact support; the remaining m may be obtained, e.g., as

$$B_{N-m+k} = [s_{N-2m+k}, \ldots, s_N]\xi_s, \quad k = 1, \ldots, m,$$

and then $B_{N-m+k(t)}(t) = 0$, for $t \le s_{N-2m+k}$.

Basis functions with compact support that span the space of natural polynomial splines with knots $s_1, \ldots, s_n, s_i \in [0,1]$ are studied in some detail in Schumaker (1981, §8.2). $n-2m$ of these basis functions are so-called B splines. These B splines are piecewise polynomials of degree $2m-1$, with $2m-2$

continuous derivatives, have exactly $2m+1$ knots, $s_l, \ldots, s_{l+2m}$, and are zero outside $[s_l, s_{l+2m}]$. It is known that (nontrivial) piecewise polynomials of the given degree and order of continuity cannot be supported on fewer knots. For equally spaced knots, with spacing h, the B splines are translated and scaled versions of the convolution of $2m$ uniform distributions on $[0, h]$. In general, they are nonnegative hill-shaped functions. They are very popular as basis functions both for their good approximation theoretic properties and their ease of computation. Simple recursion relations are available to generate them directly (see Lyche and Schumaker (1973), deBoor (1978), Schumaker (1981)). Software for generating B-spline bases given arbitrary knots s_l is publicly available (see Chapter 11).

Given basis functions $B_1, \ldots, B_N$ we now seek $f_{N,\lambda}$ of the form

$$f_{N,\lambda} = \sum_{k=1}^{N} c_k B_k$$

to minimize

$$\frac{1}{N} \sum_{i=1}^{n} \left(y_i - \sum_{k=1}^{N} x_{ik} c_k \right)^2 + \lambda \sum_{k,l=1}^{N} c_k c_l \sigma_{kl},$$

where

$$x_{ik} = \int K(t_i, s) B_k(s)\, ds$$

and

$$\sigma_{kl} = < P_1 B_k, P_1 B_l > .$$

For $\mathcal{H}_R = W_m[0,1]$,

$$\sigma_{kl} = \int_0^1 \frac{d^m}{dx^m} B_k(x) \frac{d^m}{dx^m} B_l(x)\, dx$$

and σ_{kl} will be zero if B_k and B_l have no common support. Then

$$c = c_\lambda = (X'X + n\lambda\Sigma)^{-1} X'y$$

and

$$A(\lambda) = X(X'X + n\lambda\Sigma)^{-1} X'. \tag{7.1.6}$$

Here, we really have two smoothing parameters, namely, λ and N, the number of basis functions, assuming the distribution of $s_1, \ldots, s_N$ for each N is fixed. In principle one could compute $V(N, \lambda)$ and minimize over both N and λ. We think that, to minimize errors due to numerical approximation, one wants to make N as large as is convenient or feasible on the computing equipment, or at least large enough to ensure that no "resolution" is lost at this stage, and then choose λ to minimize V.

7.2 Regression splines.

A number of authors have suggested using regression splines (i.e., $\lambda = 0$), particularly in the case $L_i f = f(t_i)$. Then N is the smoothing parameter. In this case V is particularly easy to compute, since trace $A(N) = N/n$. von Golitschek and Schumaker (1987) show, under general circumstances, that if $f \notin \mathcal{H}_N$, then it is always better to do smoothing as opposed to regression. They show that there is always some $\lambda > 0$ that is better than $\lambda = 0$ in the sense that, if $f \notin \mathcal{H}_N$, then $ET(\lambda)$ defined by

$$E\sum_{i=1}^{n}(L_i f - L_i f_\lambda)^2$$

satisfies $(d/d)\lambda ET(\lambda)|_{\lambda=0} < 0$.

Furthermore, for $L_i f = f(t_i)$, and $\mathcal{H}_R = W_m$, Agarwal and Studden (1980) have shown that the optimal N is $O(1/n^{1/2m+1}) = O(1/n^{1/5})$ for $m = 2$, say. For n of the order of, say 100, the optimal N for B-spline regression will be quite small and the estimate will not "resolve" multiple peaks that could easily be resolved by a smoothing spline. Nevertheless, arguments have been made for the use of regression splines in the case of extremely large data sets, and/or in situations where one does not expect to recover much "fine structure" in the estimates. (That is, when the true f is believed to be approximately in $\mathcal{H}_N$ for small N.) If the knots, that is the s_k, are also considered free variables, then more flexibility is available. If the knots are chosen to minimize the residual sum of squares, then trace $A(N)$ can be expected to be an underestimate of the degrees of freedom for signal in the denominator of the GCV function. Friedman and Silverman (1987) and Friedman (1989) have proposed correction factors for this.

CHAPTER 8

Fredholm Integral Equations of the First Kind

8.1 Existence of solutions, the method of regularization.

An equation of the form

$$g(t) = \int_\Omega K(t,s)f(s)\,ds, \ t \in \mathcal{T} \tag{8.1.1}$$

is known as a Fredholm integral equation of the first kind. Rewriting (8.1.1) as

$$g = \mathcal{K}f, \tag{8.1.2}$$

we have that Picard's theorem (Courant and Hilbert (1965, p.160)) says that $\mathcal{K}(\mathcal{L}_2) = \mathcal{H}_{K*K}$, where

$$K * K(u,v) = \int_\Omega K(u,s)K(v,s)\,ds.$$

Therefore a solution f in $\mathcal{L}_2$ for (8.1.2) exists provided $g \in \mathcal{H}_{K*K}$. Nashed and Wahba (1974) showed that $\mathcal{K}(\mathcal{H}_R) = \mathcal{H}_Q$, where

$$Q(u,v) = \iint_{\Omega\times\Omega} K(u,s)R(s,t)K(v,t)\,dsdt. \tag{8.1.3}$$

Therefore $g \in \mathcal{H}_Q$ ensures that there exist at least one solution $f \in \mathcal{H}_R$. (If K has a nontrivial null space in $\mathcal{H}_R$ then the solution is not unique but a unique generalized inverse can be defined; see Nashed and Wahba (1974).) The numerical solution of such equations has been a subject of intensive study over a long period of time. Until 1969 however (see Wahba (1969)) the literature concerned the case where g was presumed known exactly.

Brute force discretization of (8.1.1) to obtain a matrix equation and then solving the resultant linear system was doomed to failure because under very mild smoothness conditions on $K(\cdot,\cdot)$ adjacent rows of the matrix will become closer and closer as the discretization becomes finer and finer, and the calculation will become numerically unstable. Pretending to have infinite precision arithmetic, and infinitely accurate data, one could ask for $f_0 \in \mathcal{H}_R$ to minimize $\|P_1 f_0\|$ subject to

$$g(t_i) = \int K(t_i,s)f_0(s)\,ds, \ i = 1,2,\ldots,n.$$

In the computation of f_0, the matrix M of (1.3.9) is replaced by Σ (since $\lambda = 0$), which has ijth entry $Q(t_i, t_j)$. The condition number of Σ is the ratio of the largest to smallest eigenvalue. If the t_i's are roughly uniformly distributed, then the eigenvalues of Σ will (roughly) behave as n times the eigenvalues of Q. If K and R behave as Green's functions for kth order and $2m$th order linear differential operators, respectively, then the eigenvalues of Q will decay at the rate $\nu^{-(2m+2k)}$ and the estimate of the condition number of Σ is $O(n^{-(2m+2k)})$. If K is analytic, the eigenvalues will decay exponentially. Even with double precision on supercomputers, such an exact solution is numerically unstable for even moderate n.

Tikhonov (1963) suggested solving

$$g\left(\frac{i}{n}\right) = \int K\left(\frac{i}{n}, s\right) f(s)\, ds$$

approximately by (roughly), finding $(f(1/n), \ldots, f(n/n))$ to minimize

$$\frac{1}{n}\sum_{i=1}^{n}\left(g\left(\frac{i}{n}\right) - \frac{1}{n}\sum_{j=1}^{n} K\left(\frac{i}{n}, \frac{j}{n}\right) f\left(\frac{j}{n}\right)\right)^2$$
$$+\lambda\sum_{j=2}^{n-1}\left(f\left(\frac{j+1}{n}\right) - 2f\left(\frac{j}{n}\right) + f\left(\frac{j-1}{n}\right)\right)^2. \qquad (8.1.4)$$

He suggested choosing λ by trial and error. Phillips (1962) and Cook (1963) suggested a similar approach, and these and related methods are sometimes called Tikhonov–Phillips regularization.

The minimization of (1.3.4) was proposed in Wahba (1969) and the use of GCV to choose λ in this context appears in Wahba (1977a).

We mention only a few other references, which are directly related to the approach discussed here: Wahba (1980a, 1981a, 1981c, 1982c), Merz (1980), Crump and Seinfeld (1982), Mendelsson and Rice (1982), Nychka, Wahba, Goldfarb, and Pugh (1984), Cox and O'Sullivan (1985), O'Sullivan (1986a), Nychka and Cox (1989), Girard (1987a,b,c). See also the books of Groetsch (1984) and the books edited by Anderson, deHoog, and Lukas (1980), and Baker and Miller (1982). A recent translation from the Russian (Tikhonov and Goncharsky (1987)), gives a number of applications of Tikhonov–Phillips regularization to ill-posed problems in various fields. The discrepancy method and trial and error for choosing λ are used in the examples reported there.

8.2 Further remarks on ill-posedness.

The typical integral equation arising in practice can be very ill-posed; basically this means that very large data sets can contain a surprisingly small amount of information about the desired solution. As an example, consider Fujita's equation considered in Wahba (1979b) and references cited there. It is

$$g(t) = \int_0^{s_{\max}} \frac{\theta s e^{-\theta s t}}{[1 - e^{-\theta s}]} f(s)\, ds \qquad (8.2.1)$$

where θ is given. This equation relates optical density along a tube after centrifugation ($g(t), t =$ distance along the tube) to $f(s)$, the molecular weight distribution of the contents of the tube. One should always look at the eigenvalues $\{\lambda_{\nu n}\}$ of the problem

$$g_i = L_i f = < \eta_i, f >,$$

which consist of M ones and the $n - M$ eigenvalues of $Q_2' \Sigma Q_2$, where Σ is the $n \times n$ matrix with ijth entry $< P_1\eta_i, P_1\eta_j >$. For the example of Fujita's equation considered in Wahba (1979b), with $n = 41$ and $M = 0$, we obtained the first five eigenvalues as 1, $10^{-3.5}, 10^{-7}, 10^{-10.5}$, and 10^{-14}. The remaining eigenvalues were "computer zero." Loosely speaking, even with extremely accurate data (say to eight figures), there are still only at most three linearly independent pieces of information in the data for this problem. The "number of independent pieces of information in the data" was considered in Wahba (1980a). Let

$$y_i = < \eta_i, f > + \epsilon_i, \quad i = 1, \ldots, n$$

and let the $n \times n$ matrix with ijth entry $< \eta_i, \eta_j >$ satisfy

$$\{< \eta_i, \eta_j >\} = \Sigma = \Gamma D \Gamma'.$$

Let

$$\begin{pmatrix} \psi_1 \\ \vdots \\ \psi_n \end{pmatrix} = D^{-1/2}\Gamma' \begin{pmatrix} \eta_1 \\ \vdots \\ \eta_n \end{pmatrix},$$

and let z be the transformed data

$$z = \Gamma' y.$$

Then

$$z_\nu = \sqrt{\lambda_\nu} < \psi_\nu, f > + \tilde{\epsilon}_\nu,$$

where λ_ν is the ith diagonal entry of D and $\tilde{\epsilon} = (\tilde{\epsilon}_1, \ldots, \tilde{\epsilon}_n)' \sim \mathcal{N}(0, \sigma^2 I)$. If $\lambda_\nu < \psi_\nu, f >^2$ is large compared to σ^2, then one can obtain a good estimate of $< \psi_\nu, f >$, and, if it is not, one cannot. One might identify "the number of independent pieces of information in the data" with the number of ν's for which $\lambda_\nu / \sigma^2 >> 1$.

We note that

$$K(t, s) = \frac{\theta s e^{-\theta s t}}{[1 - e^{-\theta s}]}$$

of (8.2.1) is infinitely differentiable in t. The "smoother" $K(\cdot, \cdot)$ is, the more ill-posed the integral equation. $K(t, s) = (t - s)_+^{m-1}/(m-1)!$ corresponds to $g^{(m)} = f$ for m some positive integer. The larger m is, the more ill-posed the problem. For $m < 1$ we have Abel's equations (see Anderssen and Jakeman (1975), Nychka et al. (1984)). A plot of the relevant eigenvalues appears in Nychka et al. Abel's equations are only mildly ill-posed. The equations arising in computerized tomography (Radon transforms; see e.g., Wahba (1981c) and references cited there) are also only mildly ill-posed.

8.3 Mildly nonlinear integral equations.

Remote sensing experiments frequently involve the observation of data on mildly nonlinear functions. For example, upwelling radiation above the atmosphere is related to the atmospheric vertical temperature distribution, and this radiation is measured from sensors aboard satellites (see, e.g., Fritz et al. (1972)). It is desired to recover the vertical temperature distribution from this data for the purpose of estimating initial conditions for numerical weather prediction. With some idealizations the relation is

$$R_\nu(T) = \int_{\text{surface}}^{\text{top}} \mathcal{B}_\nu(T(x))\tau_\nu'(x)\,dx \tag{8.3.1}$$

where $T(x)$ is the temperature at vertical coordinate x along a column of the atmosphere in the line of sight of the satellite sensor, R_ν is upwelling radiance at wavenumber ν, and $\mathcal{B}_\nu$ is Planck's function

$$\mathcal{B}_\nu[T(x)] = \frac{c_1\nu^3}{e^{c_2\nu/T(x)} - 1} \tag{8.3.2}$$

where c_1 and c_2 are known physical constants and τ_ν is the transmittance (usually assumed known). The data model is

$$y_\nu = R_\nu(T) + \epsilon_\nu, \ \nu = 1, \ldots, n.$$

The following approach was proposed in O'Sullivan and Wahba (1985).

Let

$$T(x) \simeq \sum_{k=1}^{N} c_k B_k(x)$$

where the B_k are B-splines, and let

$$R_\nu(T) \simeq N_\nu(c) = \int \mathcal{B}_\nu\left(\sum_{k=1}^{N} c_k B_k(x)\right)\tau_\nu'(x)\,dx. \tag{8.3.3}$$

Find $c = (c_1, \ldots, c_N)'$ to minimize

$$\frac{1}{n}\sum_{i=1}^{n}(y_i - N_i(c))^2 + \lambda c'\Sigma c \tag{8.3.4}$$

where $\Sigma = \{\sigma_{ij}\}$, $\sigma_{ij} =< P_1B_i, P_1B_j >$. Fix λ, and use a Gauss–Newton iteration to find $c = c(\lambda)$: For $c = c^{(l)}$, the lth iterate, we have

$$N_i(c) \approx N_i(c^{(l)}) + \sum_{k=1}^{N} \left.\frac{\partial N_i}{\partial c_k}\right|_{c=c^{(l)}} (c_k - c_k^{(l)}). \tag{8.3.5}$$

Let $X^{(l)}$ be the $n \times N$ matrix with ikth entry $\partial N_i/\partial c_k|_{c=c^{(l)}}$, and let the "pseudodata" $y^{(l)}$ be

$$y^{(l)} = y - \begin{pmatrix} N_1(c^{(l)}) \\ \vdots \\ N_n(c^{(l)}) \end{pmatrix} + X^{(l)}c^{(l)}. \tag{8.3.6}$$

The minimization problem becomes: Find $c^{(l+1)}$ to minimize

$$\frac{1}{n}\|y^{(l)} - X^{(l)}c\|^2 + \lambda c'\Sigma c, \tag{8.3.7}$$

and

$$c^{(l+1)} = (X^{(l)'}X^{(l)} + n\lambda\Sigma)^{-1}X^{(l)'}y^{(l)}.$$

This iteration is run to convergence, say until $l = L = L(\lambda)$. Then the quadratic approximation to the original optimization problem (8.3.4) in the neighborhood of $c^{(L)}$ has the influence matrix $A^{(L)}(\lambda)$,

$$A^{(L)}(\lambda) = X^{(L)}(X^{(L)'}X^{(L)} + n\lambda\Sigma)^{-1}X^{(L)'} \tag{8.3.8}$$

and the GCV function can be evaluated for this λ as

$$\frac{\frac{1}{n}\,\mathrm{RSS}(\lambda)}{\frac{1}{n}\,\mathrm{Tr}\,(I - A^{(L)}(\lambda))^2},$$

and the process repeated for a new λ.

8.4 The optimal λ for loss functions other than predictive mean-square error.

The GCV estimate of $\hat{\lambda}$ has been shown to be good for estimating the λ that minimizes

$$T(\lambda) = \frac{1}{n}\sum_{i=1}^{n}(L_i f - L_i f_\lambda)^2.$$

Suppose that one is really interested in choosing λ to minimize some other loss function, for example,

$$D(\lambda) = \int (f_\lambda(t) - f(t))^2\,dt.$$

If $T(\lambda)$ and $D(\lambda)$ have to a good approximation the same minimizer, then it is sufficient to use $\hat{\lambda}$. Examples where this is so numerically appear in Craven and Wahba (1979). Suggestions for modifying the GCV function for other loss functions have been made, notably in O'Sullivan (1986a). In the comments to that paper I argued that, if the original problem is ill-conditioned, then the computation of the modified GCV function is also likely to be ill-conditioned. In any case it would be nice to know when T and D (or other loss functions) are likely to have (approximately) the same minimizer. A number of authors have provided convergence rate calculations that contribute to an answer to this question, including those in Section 4.5.

In Wahba and Wang (1987) we examined a simple family of cases that suggests the range of results that might be obtainable in general. The result is that under a range of circumstances the optimal rate of decay of λ is the same for a variety of loss functions, and under other circumstances the optimal rate of decay is different for different loss functions. We summarize the results here. We let

$$\begin{aligned} g(t) &= \int_0^1 h(t-s)f(s)\,ds, \ t \in [0,1], \\ y_i &= g\left(\frac{i}{n}\right) + \epsilon_i \end{aligned}$$

and we assumed that

$$\begin{aligned} g(t) &= \sum_{\nu=1}^{\infty} 2g_\nu \cos 2\pi\nu t, \\ h(t) &= \sum_{\nu=1}^{\infty} 2h_\nu \cos 2\pi\nu t, \\ f(t) &= \sum_{\nu=1}^{\infty} 2f_\nu \cos 2\pi\nu t, \end{aligned}$$

thus $g_\nu = h_\nu f_\nu$. f is estimated as a periodic function that integrates to zero an minimizes

$$\frac{1}{n}\sum_{i=1}^{n}\left(y_i - \int_0^1 h\left(\frac{i}{n} - s\right) f(s)\,ds\right)^2 + \frac{\lambda}{(2\pi)^{2m}}\int_0^1 (f^{(m)}(t))^2 dt.$$

Letting

$$\tilde{g}_\nu = \frac{\sqrt{2}}{n}\sum_{i=1}^{n} y_i \cos 2\pi\nu\left(\frac{i}{n}\right),$$

then, to an approximation good enough for our purposes

$$f_\lambda(s) \approx 2\sum_{\nu=1}^{n} \hat{f}_\nu \cos 2\pi\nu s$$

where

$$\hat{f}_\nu = \frac{h_\nu \tilde{g}_\nu}{h_\nu^2 + \lambda\nu^{2m}};$$

furthermore,

$$g_\lambda(t) = \int_0^1 h(t-s) f_\lambda(s)\,ds \simeq 2\sum_{\nu=1}^{n} \hat{g}_\nu \cos 2\pi\nu t$$

with

$$\hat{g}_\nu = h_\nu \hat{f}_\nu.$$

Then the mean-square error in the solution is

$$\int_0^1 (f_\lambda(s) - f(s))^2\,ds \approx \sum_{\nu=1}^{n} (\hat{f}_\nu - f_\nu)^2,$$

the mean-square prediction error is

$$\frac{1}{n}\sum_{\nu=1}^{n}\left(g_\lambda\left(\frac{i}{n}\right)-g\left(\frac{i}{n}\right)\right)^2 \approx \int_0^1 (g_\lambda(t)-g(t))^2 dt \approx \sum_{\nu=1}^{n}(\hat{g}_\nu - g_\nu)^2$$

$$= \sum_{\nu=1}^{n} h_\nu^2 (\hat{f}_\nu - f_\nu)^2,$$

and the mean-square error in the lth derivative of the solution, if it exists, is

$$\int_0^1 (f_\lambda^{(l)}(s) - f^{(l)}(s))^2 ds \approx \sum_{\nu=1}^{n}(2\pi\nu)^{2l}(\hat{f}_\nu - f_\nu)^2.$$

Now

$$\tilde{g}_\nu = \frac{\sqrt{2}}{n}\sum_{i=1}^{n}(g(\frac{i}{n}) + \epsilon_i)\cos 2\pi\nu(\frac{i}{n}) \approx g_\nu + \tilde{\epsilon}_\nu,$$

where $\tilde{\epsilon}_\nu \sim \mathcal{N}(0, \sigma^2/n)$ giving

$$E(\hat{f}_\nu - f_\nu)^2 \approx \left(\frac{\lambda\nu^{2m}}{h_\nu^2 + \lambda\nu^{2m}}\right)^2 f_\nu^2 + \frac{h_\nu^2\sigma^2}{(h_\nu^2 + \lambda\nu^{2m})^2}. \tag{8.4.1}$$

Wahba and Wang (1987) considered loss functions of the form

$$T_q(\lambda) = \sum_{\nu=1}^{n} q_\nu(\hat{f}_\nu - f_\nu)^2 \tag{8.4.2}$$

for

$$q_\nu \approx \nu^\gamma,\ f_\nu \approx \nu^{-\alpha},\ h_\nu \approx \nu^{-\beta},\ \alpha, \beta > 0. \tag{8.4.3}$$

Thus $\gamma = 0$ corresponds to mean-square solution error, $\gamma = -2\beta$ corresponds to mean-square prediction error, and $\gamma = 2m$ corresponds to $\|f - f_\lambda\|_R^2$. Substituting (8.4.3) and (8.4.1) into (8.4.2), one obtains

$$ET_q(\lambda) \approx \lambda^2 \sum_{\nu=1}^{n} \frac{\nu^{4(m+\beta)+\gamma-2\alpha}}{(1+\lambda\nu^{2(m+\beta)})^2} + \frac{\sigma^2}{n}\sum_{\nu=1}^{n}\frac{\nu^{\gamma+2\beta}}{(1+\lambda\nu^{2(m+\beta)})^2}, \tag{8.4.4}$$

and the optimal λ for the interesting combinations of α, β, γ, and m were found. We only consider the case $\beta > 0$ (which guarantees a bona fide convolution equation), $\alpha + \beta > 1$ (which guarantees that g is in a reproducing kernel space), and $m > \frac{1}{4}$. We only repeat the results for $\gamma = 0$ and -2β here. Let λ_D minimize (8.4.4) for $\gamma = 0$ (mean-square solution error (domain error)) and λ_* be the minimizer for $\gamma = -2\beta$ (mean-square prediction error). The results are:

(A) Suppose $\frac{1}{2} < \alpha \le 2m + 1$. Then

$$\lambda_* \approx \lambda_D \approx n^{-(m+\beta)/(\alpha+\beta)}$$

(B.1) Suppose $2m + 1 < \alpha$ and $\beta > (\alpha - (2m + \frac{1}{2}))$. Then

$$\lambda_* \approx \lambda_D \approx n^{-(m+\beta)/(\alpha+\beta)}$$

(B.2) Suppose $2m + 1 < \alpha$ and $\beta \leq (\alpha - (2m + \frac{1}{2}))$. Then λ_* does not (otherwise) depend on α and

$$\lambda_* \approx o(\lambda_D).$$

CHAPTER 9

Further Nonlinear Generalizations

9.1 Partial spline models in nonlinear regression.

Consider the nonlinear partial spline model

$$y_i = \psi(t_i, \theta) + f(t_i) + \epsilon_i, \quad i = 1, \ldots, n \tag{9.1.1}$$

where $\theta = (\theta_1, \ldots, \theta_q)$ is unknown, $\psi(t_i, \theta)$ is given, $f \in \mathcal{H}_R$, and $\epsilon = (\epsilon_1, \ldots, \epsilon_n)' \sim \mathcal{N}(0, \sigma^2 I)$.

We can fit this data by finding $\theta \in E^q$, $f \in \mathcal{H}$ to minimize

$$\frac{1}{n} \sum_{i=1}^{n} (y_i - \psi(t_i, \theta) - f(t_i))^2 + \lambda \|f\|^2. \tag{9.1.2}$$

Note that we have used a norm rather than a seminorm in the penalty functional in (9.1.2). Here any part of the "signal" for which there is to be no penalty should be built into ψ, to avoid hard-to-analyze aliasing when (9.1.2) is minimized using iterative methods. In most applications, f would be a smooth nuisance parameter, and testing whether or not it is zero would be a way of testing whether or not the model $\psi(t, \theta)$ is adequate.

It is easy to see that the minimizer f_λ of (9.1.2) must be of the form

$$f = \sum_{i=1}^{n} c_i R_{t_i}$$

where R_{t_i} is the representer of evaluation at t_i, in $\mathcal{H}_R$. Letting $\psi(\theta) = (\psi(t_1, \theta), \ldots, \psi(t_n, \theta))'$ and Σ be the $n \times n$ matrix with ijth entry $R(t_i, t_j)$, we have that the minimization problem of (9.1.2) becomes the following. Find $\theta \in E^q$ and $c \in E^n$ to minimize

$$\frac{1}{n} \|y - \psi(\theta) - \Sigma c\|^2 + \lambda c' \Sigma c. \tag{9.1.3}$$

For any fixed θ, we have (assuming Σ is of full rank), $c = c(\theta)$ satisfies

$$(\Sigma + n\lambda I) c(\theta) = y - \psi(\theta). \tag{9.1.4}$$

Substituting (9.1.4) into (9.1.3), (9.1.3) becomes the following. Find θ to minimize

$$\lambda(y - \psi(\theta))'(\Sigma + n\lambda I)^{-1}(y - \psi(\theta)). \tag{9.1.5}$$

The Gauss–Newton iteration goes as follows.

Let $T^{(l)}$ be the $n \times q$ matrix with $i\nu$th entry $\partial\psi(t_i, \theta)/\partial\theta_\nu|_{\theta=\theta^{(l)}}$, where $\theta^{(l)}$ is the lth iterate of θ. Expanding $\psi(\theta)$ in a first-order Taylor series gives

$$\psi(\theta) \simeq \psi(\theta^{(l)}) - T^{(l)}\theta^{(l)} + T^{(l)}\theta.$$

Letting

$$z^{(l)} = y - \psi(\theta^{(l)}) + T^{(l)}\theta^{(l)},$$

we have that $\theta^{(l+1)}$ is the minimizer of

$$(z^{(l)} - T^{(l)}\theta)'(\Sigma + n\lambda I)^{-1}(z^{(l)} - T^{(l)}\theta).$$

That is, $\theta^{(l+1)}$ satisfies

$$T^{(l)'}(\Sigma + n\lambda I)^{-1}T^{(l)}\theta^{(l+1)} = T^{(l)'}(\Sigma + n\lambda I)^{-1}z^{(l)}.$$

Letting

$$\Sigma = UDU'$$

gives

$$T^{(l)'}U(D + n\lambda I)^{-1}U'T^{(l)}\theta^{(l+1)} = T^{(l)}U(D + n\lambda I)^{-1}U'z^{(l)},$$

so that the same $n \times n$ matrix decomposition can be used for all iterations and values of λ.

For fixed λ, the iteration is carried to convergence, $l = L = L(\lambda)$, say, and the solution $(\theta_\lambda, c_\lambda)$ is the solution to the linearized problem

$$\frac{1}{n}\|z^{(L)} - T^{(L)}\theta - \Sigma c\|^2 + \lambda c'\Sigma c, \tag{9.1.6}$$

for which the influence matrix $A(\lambda) = A^{(L)}(\lambda)$ is given by the familiar formula

$$I - A^{(L)}(\lambda) = n\lambda Q_2^{(L)}(Q_2^{(L)'}\Sigma Q_2^{(L)} + n\lambda I)^{-1}Q_2^{(L)'} \tag{9.1.7}$$

where

$$T^{(L)} = (Q_1^{(L)} : Q_2^{(L)})\begin{pmatrix} R^{(L)} \\ 0 \end{pmatrix}.$$

One has $n\lambda c = (I - A^{(L)})z^{(L)}$, and the GCV function can be defined as

$$V(\lambda) = \frac{\frac{1}{n}\|(I - A^{(L)}(\lambda))z^{(L)}\|^2}{(\frac{1}{n}\mathrm{Tr}\,(I - A^{(L)}(\lambda)))^2}.$$

The matrix $A^{(L)}$ has q eigenvalues that are one, thus this formula is assigning q degrees of freedom for signal to the estimation of $(\theta_1, \ldots, \theta_q)$. It has been noted by many authors (see the comments in Friedman and Silverman (1989)) that when q parameters enter nonlinearly q may not be the real equivalent degrees of freedom. It is an open question whether a correction needs to be made here, and what modifications, if any, to the hypothesis testing procedure in Section 6 when the null space is not a linear space. See the recent book by Bates and Watts (1988) for more on nonlinear regression.

9.2 Penalized GLIM models.

Suppose

$$y_i \sim \text{Binomial } (1, p(t_i)), \qquad i = 1, \ldots, n, \tag{9.2.1}$$

and let the logit $f(t)$ be $f(t) = \log[p(t)/(1-p(t))]$, where f is assumed to be in $\mathcal{H}_R$, and it is desired to estimate f. The negative log-likelihood $\mathcal{L}(y)$ of the data is

$$\mathcal{L}(y) = -\sum_{i=1}^{n} (y_i \log p(t_i) + (1-y_i) \log (1-p(t_i))) \tag{9.2.2}$$

and, since $p = e^f/(1+e^f)$,

$$\mathcal{L}(y) = \sum_{i=1}^{n} (\log (1+e^{f(t_i)}) - y_i f(t_i)) = Q(y, f), \text{ say}. \tag{9.2.3}$$

McCullagh and Nelder (1983) in their book on GLIM (generalized linear models) suggest assuming that f is a parametric function (for example, $f(t) = \theta_1 + \theta_2 t$), and estimating the parameters by minimizing $Q(y, f) = Q(y, \theta)$. O'Sullivan (1983) and O'Sullivan, Yandell, and Raynor (1986) considered the estimation of f by supposing that $f \in \mathcal{H}_R$ and finding f to minimize the penalized log-likelihood

$$I_\lambda(y, f) = Q(y, f) + \lambda \|P_1 f\|^2, \tag{9.2.4}$$

and extended GCV to this setup.

Penalized likelihood estimates with various penalty functionals have been proposed by a number of authors. (See the references in O'Sullivan (1983); we note only the work of Silverman (1978) and Tapia and Thompson (1978) where the penalty functionals are seminorms in reproducing kernel spaces. See also Leonard (1982).)

If (9.2.1) is replaced by

$$y_i \sim \text{Poisson } (\Lambda(t_i))$$

and $f(t) = \log \Lambda(t)$, we have

$$Q(y, f) = \sum_{i=1}^{n} \left\{ e^{f(t_i)} - y_i f(t_i) + \log(y_i!) \right\}. \tag{9.2.5}$$

Of course $y_i \sim \mathcal{N}(f(t_i), \sigma^2)$ is the model we have been considering all along, and the setup we are discussing works whenever y_i has an exponential density of the form

$$p(y_i) = e^{-[\{b(f(t_i)) - y_i f(t_i)\}/a_i] + c(y_i)} \tag{9.2.6}$$

where a_i, b, and c are given.

Here we note for further reference that $Ey_i = b'(f(t_i))$ and var $y_i = b''(f(t_i))a_i$, and below we will let $a_i = 1$.

Approximating f by a suitable basis function representation in $\mathcal{H}_R$, we have

$$f \simeq \sum_{k=1}^{N} c_k B_k$$

and we need to find $c = (c_1, \ldots, c_N)'$ to minimize

$$I_\lambda(c) = \sum_{i=1}^{n} b\left(\sum_{k=1}^{N} c_k B_k(t_i)\right) - y_i \left(\sum_{k=1}^{N} c_k B_k(t_i)\right)$$

$$+ \lambda \sum_{k,k'=1}^{N} c_k c_{k'} < P_1 B_k, P_1 B_{k'} > . \tag{9.2.7}$$

Using a Newton–Raphson iteration this problem can be solved iteratively, and at the last step of the iteration one can obtain an approximating quadratic problem, from which one can extract a GCV function.

The second-order Taylor expansion of $I_\lambda(c)$ for the lth iterate $c^{(l)}$ is

$$I_\lambda(c) \simeq I_\lambda(c^{(l)}) + \nabla I_\lambda(c - c^{(l)}) + \frac{1}{2}(c - c^{(l)})' \nabla^2 I_\lambda(c - c^{(l)}) \tag{9.2.8}$$

where the gradient ∇I_λ is given by

$$\nabla I_\lambda = \left(\frac{\partial I_\lambda}{\partial c_1}, \ldots, \frac{\partial I_\lambda}{\partial c_N}\right)\Bigg|_{c=c^{(l)}} \tag{9.2.9}$$

and the Hessian $\nabla^2 I_\lambda$ is the $N \times N$ matrix with jkth entry

$$\{\nabla^2 I_\lambda\}_{jk} = \frac{\partial^2 I_\lambda}{\partial c_j \partial c_k}\Bigg|_{c=c^{(l)}} . \tag{9.2.10}$$

Then $c = c^{(l+1)}$, the minimizer of (9.2.8), is given by

$$c^{(l+1)} = c^{(l)} - (\nabla^2 I_\lambda)^{-1} \nabla I'_\lambda. \tag{9.2.11}$$

Letting X be the $n \times N$ matrix with ikth entry $B_k(t_i)$ and Σ be the $N \times N$ matrix with kk'th entry $< P_1 B_k, P_1 B_{k'} >$, we have that (9.2.7) becomes

$$I_\lambda(c) = \sum_{i=1}^{n} b((Xc)_i) - y'Xc + \lambda c'\Sigma c,$$

where $(Xc)_i$ is the ith entry of the vector Xc. We have

$$\nabla I'_\lambda = X'(\mu(c) - y) + 2\lambda\Sigma c \tag{9.2.12}$$

where $\mu(c) = (\mu_1(c), \ldots, \mu_n(c))'$ with $\mu_i(c) = b'(f(t_i)) = b'((Xc)_i)$; and

$$\nabla^2 I_\lambda = X'D(c)X + 2\lambda\Sigma \tag{9.2.13}$$

where $D(c)$ is the $n \times n$ diagonal matrix with iith entry $b''(f(t_i)) = b''((Xc)_i)$.

Substituting (9.2.12) and (9.2.13) into (9.2.11) gives the Newton–Raphson update

$$c^{(l+1)} = c^{(l)} - (X'D(c^{(l)})X + 2\lambda\Sigma)^{-1}(-X'(y - \mu(c^{(l)})) + 2\lambda\Sigma c^{(l)}) \quad (9.2.14)$$

$$= (X'D(c^{(l)})X + 2\lambda\Sigma)^{-1}X'D^{1/2}(c^{(l)})z^{(l)} \quad (9.2.15)$$

where the pseudodata $z^{(l)}$ is

$$z^{(l)} = D^{-1/2}(c^{(l)})(y - \mu(c^{(l)})) + D^{1/2}(c^{(l)})Xc^{(l)}. \quad (9.2.16)$$

Then $c^{(l+1)}$ is the minimizer of

$$\frac{1}{2}\|z^{(l)} - D^{1/2}(c^{(l)})Xc\|^2 + \lambda c'\Sigma c. \quad (9.2.17)$$

The predicted value $\hat{z}^{(l)} = D^{1/2}Xc$ of $z^{(l)}$ is related to $z^{(l)}$ by

$$\hat{z}^{(l)} = A(\lambda)z^{(l)}$$

where

$$A(\lambda) = D^{1/2}(c^{(l)})X(X'D(c^{(l)})X + \lambda\Sigma)^{-1}X'D^{1/2}(c^{(l)}). \quad (9.2.18)$$

Running (9.2.14) to convergence, (9.2.17) at convergence becomes

$$\frac{1}{2}\|D^{-1/2}(c)(y - \mu(c))\|^2 + \lambda c'\Sigma c \quad (9.2.19)$$

and letting $w = D^{-1/2}(c)y$ and $\hat{w} = D^{-1/2}\mu(c)$, it is seen that

$$\frac{\partial \hat{w}_i}{\partial w_j} \simeq (A(\lambda))_{ij}$$

resulting in the GCV function

$$V(\lambda) = \frac{\|D^{-1/2}(c)(y - \mu(c))\|^2}{(\text{Tr }(I - A(\lambda)))^2}$$

evaluated at the converged value of c.

Properties of these estimates are discussed in O'Sullivan (1983), Cox and O'Sullivan (1989a,b), and Gu (1989a). The method has been extended to the Cox proportional hazards model and other applications by O'Sullivan (1986b, 1988b).

9.3 Estimation of the log-likelihood ratio.

Suppose one is going to draw a sample of n_1 observations from population 1 with density $h_1(t)$, $t \in \mathcal{T}$, and a sample of n_2 observations from population 2 with density $h_2(t)$, and it is desired to estimate $f(t) = \log\,(h_1(t)/h_2(t))$, the log-likelihood ratio. Without loss of generality we will suppose $n_1 = n_2 = n/2$. (Details for removing this limitation may be found in Villalobos and Wahba (1987).) Suppose the n observations are labeled $t_1, \ldots, t_n$, and with each observation is attached a tag y_i, $i = 1, \ldots, n$ that is 1 if t_i came from population 1, and 0 if t_i came from population 2.

Given that an observation has value t_i, the conditional probability that its tag y_i is 1, is $h_1(t_i)/(h_1(t_i) + h_2(t_i)) = p(t_i)$, and we have that $f(t) = \log(p(t)/(1 - p(t)) = \log(h_1(t)/h_2(t))$. f can be estimated by minimizing $Q(y, f) + \lambda \|P_1 f\|^2$, where Q is given by (9.2.3). This way of looking at log likelihood estimation is due to Silverman (1978).

Note that if h_1 and h_2 are d-variate normal densities, $(t = (x_1, \ldots, x_d))$, then f is quadratic in $x_1, \cdots, x_d$ and will be in the null space of the thin plate penalty functional for $m = 3$ (provided $6 - d > 0$; see Section 2.4). Thus, if h_1 and h_2 are believed to be "close" to multivariate normal, then this penalty functional is a natural one (see Silverman (1982)).

9.4 Linear inequality constraints.

Suppose we observe

$$y_i = L_i f + \epsilon_i, \tag{9.4.1}$$

and it is known a priori that $f \in \mathcal{C} \subset \mathcal{H}_R$ where $\mathcal{C}$ is a closed convex set.

We want to find $f \in \mathcal{H}_R$ to minimize

$$\frac{1}{n}\sum_{i=1}^{n}(y_i - L_i f)^2 + \lambda \|P_1 f\|^2 \tag{9.4.2}$$

subject to $f \in \mathcal{C}$. Since any closed convex set can be characterized as the intersection of a family of half planes, we can write

$$\mathcal{C} = \{f : < \chi_s, f > \geq \alpha(s), \quad s \in \mathcal{S}\},$$

for some family $\{\chi_s, \ s \in \mathcal{S}\}$. Frequently, we can approximate $\mathcal{C}$ by $\mathcal{C}_L$,

$$\mathcal{C}_L = \{f : < \chi_s, f > \geq \alpha(s), \ s = s_1, \ldots, s_L\},$$

where $\chi_{s_1}, \ldots, \chi_{s_L}$ is a discrete approximation to $\{\chi_s, \ s \in \mathcal{S}\}$. For example, if $\mathcal{C} = \{f : \ f(t) \geq 0, \ t \in \mathcal{T}\}$, then we have $\{\chi_s, \ s \in \mathcal{S}\} = \{R_s, \ s \in \mathcal{T}\}$, and if $\mathcal{T} = [0, 1]$, we may approximate $\mathcal{C}$ by $\mathcal{C}_L = \{R_{1/L}, R_{2/L}, \ldots, R_{L/L}\}$. If $\mathcal{C}_L$ is a good approximation to $\mathcal{C}$, one may frequently find after minimizing (9.4.2) subject to $f \in \mathcal{C}_L$ that the result is actually in $\mathcal{C}$. Letting η_i be the representer of L_i and $\xi_i = P_1 \eta_i$, and letting $\rho_j = P_1 \chi_{s_j}$, it is known from Kimeldorf and Wahba (1971), that if

$$\frac{1}{n}\sum_{i=1}^{n}(y_i - L_i f)^2$$

has a unique minimizer in $\mathcal{H}_0$, then (9.4.2) has a unique minimizer in $\mathcal{C}_L$, and it must have a representation

$$\sum_{i=1}^{n} c_i \xi_i + \sum_{j=1}^{L} b_j \rho_j + \sum_{\nu=1}^{M} d_\nu \phi_\nu$$

for some coefficient vectors $a = (c' : b')'$, and d. The coefficients a and d are found by solving the following quadratic programming problem. Find $a \in E^{n+L}$, $d \in E^d$ to minimize

$$\|\Sigma_1 a + T_1 d - y\|^2 + n\lambda a' \Sigma a \tag{9.4.3}$$

subject to

$$\Sigma_2 a + T_2 d \geq \alpha \tag{9.4.4}$$

where

$$\Sigma_1 = (\Sigma_{11} : \Sigma_{12}), \tag{9.4.5}$$

$$\Sigma_2 = (\Sigma_{21} : \Sigma_{22}), \tag{9.4.6}$$

$$\Sigma = \begin{pmatrix} \Sigma_1 \\ \cdots \\ \Sigma_2 \end{pmatrix} \tag{9.4.7}$$

and the Σ_{ij} and T_i are given in Table 9.1 ($\Sigma_{12} = \Sigma'_{21}$), and $\alpha = (\alpha(s_1), \ldots, \alpha(s_L))'$.

TABLE 9.1
Definitions of Σ_{ij} and T_i.

Matrix	Dimension	ijth entry
Σ_{11}	$n \times n$	$< \xi_i, \xi_j >$
Σ_{12}	$n \times L$	$< \xi_i, \rho_j >$
Σ_{22}	$L \times L$	$< \rho_i, \rho_j >$
T_1	$n \times M$	$< \eta_i, \phi_j >$
T_2	$L \times M$	$< \chi_{s_i}, \phi_j >$

A GCV function can be obtained for constrained problems via the "leaving-out-one" lemma of Section 4.2.

Let $f_\lambda^{[k]}$ be the minimizer in $\mathcal{C}_L$ of

$$\frac{1}{n} \sum_{\substack{i=1 \\ i \neq k}}^{n} (y_i - L_i f)^2 + \lambda \|P_1 f\|^2$$

(supposed unique) and let

$$a_{kk}^*(\lambda, \delta) = \frac{L_k f_\lambda[k, \delta] - L_k f_\lambda}{\delta} \tag{9.4.8}$$

where $f_\lambda[k,\delta]$ is the minimizer of (9.4.2) in $\mathcal{C}_L$ with y_k replaced by $y_k+\delta$. If there are no active constraints, then $L_k f_\lambda$ is linear in the components of y and

$$a_{kk}^*(\lambda,\delta) = \frac{\partial L_k f_\lambda}{\partial y_k} = a_{kk}(\lambda), \tag{9.4.9}$$

where $a_{kk}(\lambda)$ is the kkth entry of the influence matrix of (1.3.23). From Theorem 4.2.1, we have that the ordinary cross-validation function $V_0(\lambda)$ satisfies

$$V_0(\lambda) \equiv \frac{1}{n}\sum_{i=1}^n (y_k - L_k f_\lambda^{[k]})^2 \equiv \frac{1}{n}\sum_{i=1}^n (y_k - L_k f_\lambda)^2/(1-a_{kk}^*(\lambda,\delta_k)) \tag{9.4.10}$$

where $\delta_k = L_k f_\lambda^{[k]} - y_k$. By analogy with the linear case, the GCV function is defined as

$$V(\lambda) = \frac{\frac{1}{n}\sum_{i=1}^n (y_i - L_i f_\lambda)^2}{(1-\frac{1}{n}\sum_{k=1}^n a_{kk}^*(\lambda,\delta_k))^2}. \tag{9.4.11}$$

To evaluate $V(\lambda)$ for a single value of λ we need to solve n quadratic programming problems in $n+L-M$ variables. To avoid this it is suggested in Wahba (1982c) that the approximate GCV function

$$V_{\text{app}}(\lambda) = \frac{\frac{1}{n}\sum_{i=1}^n (y_i - L_i f_\lambda)^2}{(1-\frac{1}{n}\sum_{k=1}^n a_{kk}(\lambda))^2} \tag{9.4.12}$$

where

$$a_{kk}(\lambda) = \left.\frac{\partial L_k f_\lambda}{\partial y_k}\right|_y \tag{9.4.13}$$

be used. The right-hand side of (9.4.13) is well defined and continuous in λ except at boundaries when a constraint changes from active to inactive or vice versa.

We can obtain $\partial L_k f_\lambda/\partial y_k$ for the constrained problem by examining an approximating quadratic minimization problem. It is not hard to see that the approximating problem is found as follows. Fix λ, and solve the quadratic programming problem of (9.4.3) and (9.4.4). Find all the active constraints (suppose there are l). Now let $\tilde{\Sigma}_1, \tilde{\Sigma}_2, \tilde{\Sigma}, \tilde{T}_1$, and $\tilde{T}_2$, and $\tilde{\alpha}$, be defined as the corresponding elements in (9.4.5), (9.4.6), (9.4.7), and Table 9.1 with all rows and/or columns corresponding to inactive constraints deleted. Then c, d, and the nonzero values of b in the solution to quadratic programming problems of (9.4.3) and (9.4.4) are given by the solution to the following equality constrained minimization problem. Find $\tilde{a}, \tilde{d}$ to minimize

$$\|\tilde{\Sigma}_1\tilde{a} + \tilde{T}_1\tilde{d} - y\|^2 + n\lambda\tilde{a}'\tilde{\Sigma}\tilde{a} \tag{9.4.14}$$

subject to

$$\tilde{\Sigma}_2\tilde{a} + T_2\tilde{d} = \tilde{\alpha}. \tag{9.4.15}$$

To write the solution to this minimization problem quickly, let

$$W_\xi = \begin{pmatrix} I_{n\times n} & O_{n\times l} \\ O_{l\times n} & \xi I_{l\times l} \end{pmatrix}$$

and consider the following minimization problem. Find $\tilde{a}$ and d to minimize

$$(\tilde{\Sigma}\tilde{a} + \tilde{T}d - \tilde{y})' W_\xi^{-1} (\tilde{\Sigma}\tilde{a} + \tilde{T}d - \tilde{y}) + n\lambda \tilde{a}'\tilde{\Sigma}\tilde{a}, \tag{9.4.16}$$

where

$$\tilde{y} = \begin{pmatrix} y \\ \tilde{\alpha} \end{pmatrix}.$$

It is not hard to see that the minimizer of (9.4.16) satisfies

$$(\tilde{\Sigma} + n\lambda W_\xi)\tilde{a} + \tilde{T}d = \tilde{y}, \tag{9.4.17}$$

$$\tilde{T}'\tilde{a} = 0, \tag{9.4.18}$$

for any W_ξ with $\xi > 0$, and if we let $\xi \to 0$ we get the minimizer of (9.4.14) subject to (9.4.15).

Let the QR decomposition of $\tilde{T}$ be

$$\tilde{T} = (\tilde{Q}_1 : \tilde{Q}_2) \begin{pmatrix} \tilde{R} \\ 0 \end{pmatrix}$$

and letting $W = W_0$, we can derive, following the arguments leading to (1.3.19),

$$\tilde{a} = \tilde{Q}_2(\tilde{Q}_2'(\tilde{\Sigma} + n\lambda W)\tilde{Q}_2)^{-1}\tilde{Q}_2'\tilde{y}. \tag{9.4.19}$$

Now

$$\hat{\tilde{y}} \equiv \begin{pmatrix} \hat{y} \\ \tilde{\alpha} \end{pmatrix} = \tilde{\Sigma}\tilde{a} + \tilde{T}d, \tag{9.4.20}$$

and subtracting (9.4.20) from (9.4.17) gives

$$\begin{pmatrix} y - \hat{y} \\ 0 \end{pmatrix} = n\lambda W\tilde{a} = n\lambda W\tilde{Q}_2(\tilde{Q}_2'(\tilde{\Sigma} + n\lambda W)\tilde{Q}_2)^{-1}\tilde{Q}_2'\tilde{y}, \tag{9.4.21}$$

so for $j = 1, \ldots, n$, the jjth entry in the matrix on the right of (9.4.21) is $1 - (\partial L_j f_\lambda / \partial y_j)|_y$. Thus

$$\sum_{j=1}^{n} a_{jj}(\lambda) = n - n\lambda \text{ tr } \Delta\,(\Phi + n\lambda\Delta)^{-1}$$

where

$$\Delta = \tilde{Q}_2' W \tilde{Q}_2$$

and

$$\Phi = \tilde{Q}_2'\tilde{\Sigma}\tilde{Q}_2;$$

furthermore,

$$\text{tr } \Delta (\Phi + n\lambda\Delta)^{-1} = \sum_{j=1}^{n+l-M} \frac{w_j}{1 + n\lambda w_j}$$

where $w_1, \ldots, w_{n+l-M}$ are the eigenvalues of the real symmetric eigenvalue problem

$$\Delta u_j = w_j \Phi u_j, \quad j = 1, \ldots, n+l-M.$$

These arguments are from Villalobos and Wahba (1987). (The derivations there are modified to apply directly to thin-plate splines.) A numerical strategy for carrying out the computation of $V_{\text{app}}(\lambda)$ is given there. It uses an "active set" algorithm of Gill et al. (1982) for the quadratic optimization problem. This type of algorithm is known to converge rapidly when a good starting guess is available for the active constraint set. If the set of active constraints changes slowly with λ, then a good guess for the active set for an updated λ is the active set for the preceding λ. The unconstrained problem is solved to obtain a starting guess. Discontinuities in $V_{\text{app}}(\lambda)$ as the active constraint set changed were evident in the examples tried in Villalobos and Wahba (1987), but were not a practical problem. Another recent work on inequality constraints is Elfving and Andersson (1986).

9.5 Inequality constraints in ill-posed problems.

In the solution of Fredholm integral equations of the first kind, if the number of linearly independent pieces of information in the data (see Section 8.2) is small, the imposition of known linear inequality constraints may add crucial missing information. A numerical experiment that illustrates this fact was performed in Wahba (1982c) and here we show two examples from that study. Data were generated according to the model

$$y_i = \int_0^1 k(\frac{i}{n} - u) f(u) du + \epsilon_i, \quad i = 1, \cdots, n \tag{9.5.1}$$

with $n = 64$, $\epsilon_i \sim \mathcal{N}(0, \sigma^2)$ and $\sigma = .05$. The periodic convolution kernel k is shown in Figure 9.1. The two example f's from that paper are the solid black lines in Figures 9.2(a) and 9.3(a). (The explicit formulae may be found in Wahba (1982c).) The dashed curves in Figures 9.2(a) and 9.3(a) are $g(x) = \int_0^1 k(x-u) f(u) du$, and the circles are the data y_i. f was estimated as the minimizer of

$$\frac{1}{n} \sum_{i=1}^{n} (y_i - \int_0^1 k\left(\frac{i}{n} - u\right) f(u) du)^2 + \lambda \int_0^1 (f''(u))^2 du \tag{9.5.2}$$

in a 64-dimensional space of sines and cosines, using GCV to estimate λ. The numerical problem here is much simplified over that in Section 9.4 due to the periodic nature of the problem and the equally spaced data. The estimates are given as the finely dashed curves marked $f_{\hat{\lambda}}$ in Figures 9.2(b) and 9.3(b).

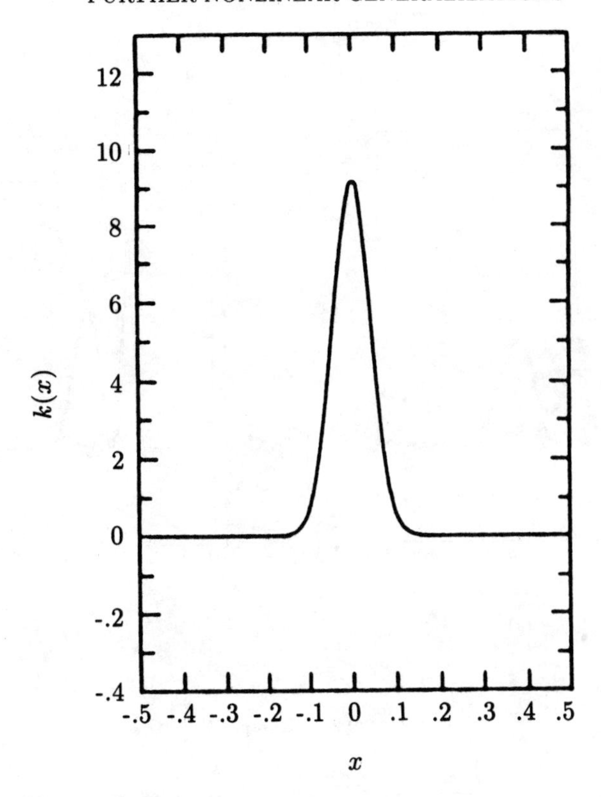

FIG. 9.1. *The convolution kernel.*

Then (9.5.2) was minimized in the same space of sines and cosines, subject to $f(\frac{i}{n}) \geq 0$, $i = 1, \ldots, n$, and V_{app} of (9.4.12) used to choose λ. The result is the coarsely dashed curves in the same figures, marked $f^c_{\hat{\lambda}_c}$. It can be seen that the imposition of the positivity constraints reduces the erroneous side lobes in the solution as well as improves the resolution of the peaks. We remark that although in theory there are 64 strictly positive eigenvalues, in this problem the ratio $(\lambda_{42}/\lambda_1)^{1/2}$ was 10^{-7}.

9.6 Constrained nonlinear optimization with basis functions.

Let

$$y_i = N_i f + \epsilon_i \tag{9.6.1}$$

where N_i is a nonlinear functional, and suppose it is known that

$$< \chi_s, f > \geq \alpha(s), \qquad s \in \mathcal{S}. \tag{9.6.2}$$

Approximating f by

$$f \simeq \sum_{k=1}^{N} c_k B_k \tag{9.6.3}$$

and $\mathcal{S}$ by $\{s_1, \ldots, s_J\}$, we seek to find c to minimize

$$\frac{1}{n}\sum_{i=1}^{n}(y_i - N_i(c))^2 + \lambda c'\Sigma c \tag{9.6.4}$$

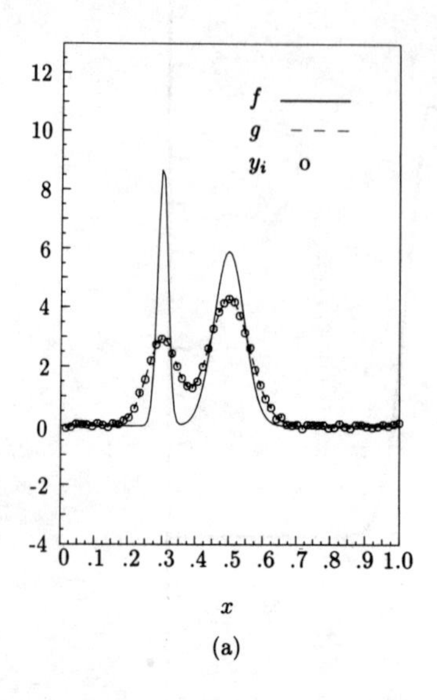

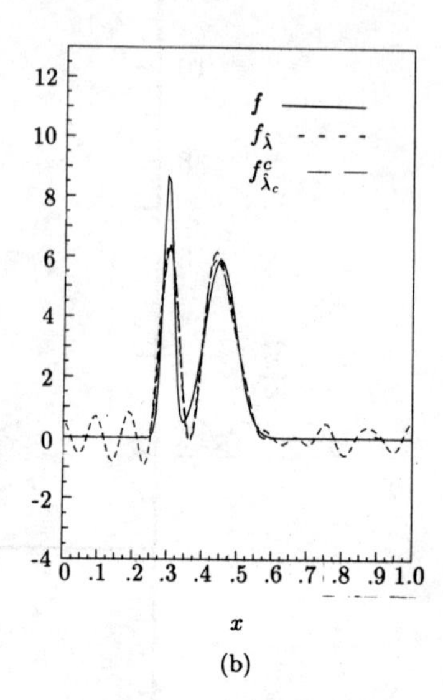

FIG. 9.2. *Example* 1.

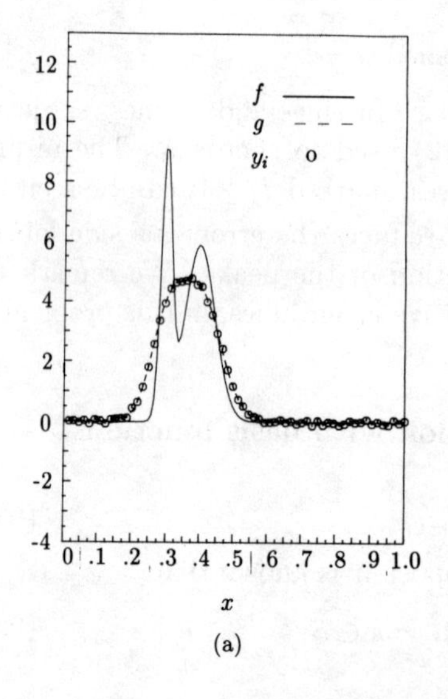

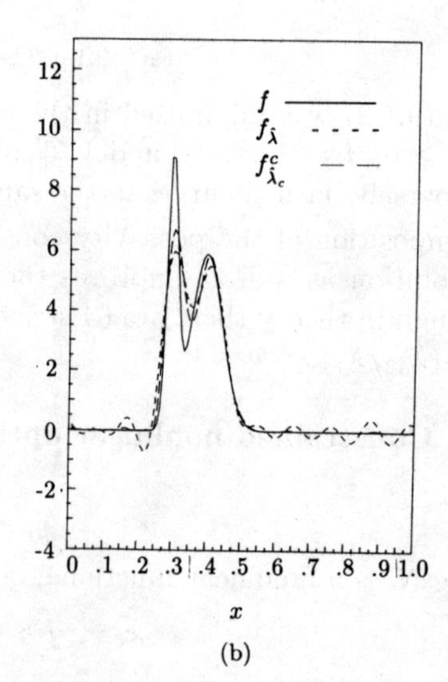

FIG. 9.3. *Example* 2.

subject to

$$\sum_{k=1}^{N} c_k < \chi_{s_r}, B_k > \geq \alpha(s_r), \qquad r = 1, \ldots, J. \tag{9.6.5}$$

Here, Σ is the $N \times N$ matrix with ijth entry $< P_1 B_i, P_1 B_j >$, as in Section 7.1. Letting

$$N_i(c) \simeq N_i(c^{(l)}) + \sum_{k=1}^{N} \frac{\partial N_i}{\partial c_k}(c_k - c_k^{(l)})$$

as in (8.3.5), and letting $X^{(l)}$ be the $n \times N$ matrix with $i\,k$th entry $\partial N_i / \partial c_k |_{c=c^{(l)}}$, and letting

$$y^{(l)} = y - \begin{pmatrix} N_1(c^{(l)}) \\ \vdots \\ N_n(c^{(l)}) \end{pmatrix} + X^{(l)} c^{(l)}$$

be as in (8.3.5)–(8.3.7), at the lth step of an iterative solution we have that the problem is to find c to minimize

$$\frac{1}{n} \| y^{(l)} - X^{(l)} c \|^2 + \lambda c' \Sigma c$$

subject to

$$Cc \geq \alpha$$

where C is the $J \times N$ matrix with rkth entry $< \chi_{s_r}, B_k >$, and $\alpha = (\alpha(s_1), \ldots, \alpha(s_J))'$. In principle at least, this problem can be iterated to convergence for fixed λ, and V_{app} for constrained functions can be evaluated. Here the influence matrix for V_{app} is

$$A^{(L)}(\lambda) = X^{(L)} F' (F X^{(L)'} X^{(L)} F' + n\lambda F \Sigma F')^{-1} F X^{(L)'}$$

where $X^{(L)}$ is the converged value of $X^{(l)}$, and, if there are J' active constraints and $C^{(L)}$ is the $J' \times N$ submatrix of C corresponding to these J' constraints, then F is any $N - J' \times N$ matrix with $F'F = I_{N-J'}$ and $FC^{(L)} = 0_{N-J' \times J}$.

9.7 System identification.

The key idea in this section (equation (9.7.14)), which allows the use of GCV in the system identification problem, has been adapted from O'Sullivan (1986a). Kravaris and Seinfeld (1985) have proposed the method of (9.7.4) below that, adopting the nomenclature of the field, might be called the penalized output least squares method. Another important recent reference is O'Sullivan (1987b), where convergence properties of the method are discussed.

The dynamic flow of fluid through a porous medium is modeled by a diffusion equation

$$\frac{\partial u(\mathbf{x}, t)}{\partial t} - \frac{\partial}{\partial \mathbf{x}} \left\{ \rho(\mathbf{x}) \frac{\partial}{\partial \mathbf{x}} u(\mathbf{x}, t) \right\} = q(\mathbf{x}, t), \;\; \mathbf{x} \in \Omega, t \in [t_{\min}, t_{\max}] \tag{9.7.1}$$

subject to prescribed initial and boundary conditions, for example, $u(\mathbf{x},0) = u_0(\mathbf{x})$ (initial condition) and $\partial u/\partial w = 0$ where w is the direction normal to the boundary. Here, if $\mathbf{x} = (x_1,\ldots,x_d)$ then $\partial/\partial\mathbf{x} = \sum_{\alpha=1}^{d} \partial/\partial x_\alpha$. Here u is, say, pressure, q represents a forcing function (injection of fluid into the region), and ρ is the transmittivity or permeability of the medium. If u_0 and q are known exactly, then for fixed ρ in some appropriate class ($\rho(\mathbf{x}) > 0$, in particular), u is determined (implicitly) as a function of ρ. Typically, ρ must be nonnegative to be physically meaningful, and such that there is measurable flow.

The practical problem is, given measurements

$$y_{ij} = u(\mathbf{x}(i), t_j, \rho) + \epsilon_{ij} \tag{9.7.2}$$

on u, the initial boundary functions, and q, estimate ρ.

We remark that if $\partial/\partial\mathbf{x}u(\mathbf{x},t)$ is zero for $\mathbf{x}$ in some region $\Omega_0 \subset \Omega$, all t, then there is no information in the experiment concerning $\rho(\mathbf{x})$ for $x \in \Omega_0$. Although the algorithm below may provide an estimate for $\rho(\mathbf{x})$ for $x \in \Omega_0$, in this case the information is coming from the prior, and not the experiment.

This is an extremely important practical problem; see, e.g., the references in O'Sullivan (1986a) and Kravaris and Seinfeld (1985). Deveaux and Steele (1989) study a somewhat different but related inverse problem.

The problem will be solved approximately in the span of a suitable set of N basis functions

$$\rho(\mathbf{x}) = \sum_{k=1}^{N} c_k B_k(\mathbf{x}),$$

and since ρ must be nonnegative, we put a sufficiently large number of linear inequality constraints on $c = (c_1,\ldots,c_N)$, that is,

$$\sum_{k=1}^{N} c_k B_k(\mathbf{x}) \geq 0 \tag{9.7.3}$$

for $\mathbf{x}$ in some finite set, so that the estimate is positive. If stronger information than just positivity is known, then it should be used. We seek to find c subject to (9.7.3) to minimize

$$\sum_{ij} (y_{ij} - u(\mathbf{x}(i), t_j, c))^2 + \lambda c'\Sigma c, \tag{9.7.4}$$

where $c'\Sigma c = \|P_1\rho\|^2$. For the moment we suppose that u_0 and q are known exactly. Then

$$u(\mathbf{x}(i), t_j, \rho) \simeq u(\mathbf{x}(i), t_j; c)$$

is a nonlinear functional of c, but only defined implicitly. If $u(\mathbf{x}(i), t_j; c)$ could be linearized about some reasonable starting guess

$$\rho_0(\mathbf{x}) = \sum_{k=1}^{N} c_k^{(0)} B_k(\mathbf{x})$$

then the methods of Section 9.6 could be used to numerically find the minimizing c_λ and to choose λ by GCV.

Given a guess $c^{(l)}$ for c, we would like to be able to linearize about $c^{(l)}$,

$$u(\mathbf{x}(i), t_j; c) \simeq u(\mathbf{x}(i), t_j; c^{(l)}) + \sum_k X_{ijk}(c_k - c_k^{(l)}), \tag{9.7.5}$$

where

$$X_{ijk} = \frac{\partial u}{\partial c_k}(\mathbf{x}(i), t_j; c)\bigg|_{c=c^{(l)}}. \tag{9.7.6}$$

If this could be done, then c and λ could be determined, at least in principle, via the constrained Gauss–Newton iteration and the GCV procedure described in Section 9.6.

Let

$$L_c = \frac{\partial}{\partial t} - \frac{\partial}{\partial \mathbf{x}}\left\{\sum_{k=1}^N c_k B_k(\mathbf{x})\frac{\partial}{\partial \mathbf{x}}\right\}, \tag{9.7.7}$$

let

$$\mathcal{B} = \{u: \ u \text{ satisfies the given initial and boundary conditions,}\}$$

$$\mathcal{B}_0 = \{u: \ u \text{ satisfies homogeneous initial and boundary conditions,}\}$$

and let

$$\delta_k = (0, \cdots, 0, \delta, 0, \cdots, 0), \ \ \delta \text{ in the } k\text{th position.}$$

Let u_c be the solution to

$$L_c u_c = q, \qquad u_c \in \mathcal{B}, \tag{9.7.8}$$

let $u_{c+\delta_k}$ be the solution to

$$L_{c+\delta_k} u_{c+\delta_k} = q, \qquad u_{c+\delta_k} \in \mathcal{B}, \tag{9.7.9}$$

and let

$$h_{c,k}(\delta) = \frac{u_{c+\delta_k} - u_c}{\delta}. \tag{9.7.10}$$

Observe that

$$L_{c+\delta_k} = L_c - \delta\frac{\partial}{\partial \mathbf{x}}B_k(\mathbf{x})\frac{\partial}{\partial \mathbf{x}}; \tag{9.7.11}$$

then substituting (9.7.9) into (9.7.10) gives

$$\left(L_c - \delta\frac{\partial}{\partial \mathbf{x}}B_k(\mathbf{x})\frac{\partial}{\partial \mathbf{x}}\right)(u_c + \delta h_{c,k}(\delta)) = q, \tag{9.7.12}$$

$$u_c + \delta h_{c,k}(\delta) \in \mathcal{B}.$$

Subtracting (9.7.8) from (9.7.12) and dividing through by δ gives

$$L_c h_{c,k}(\delta) = \frac{\partial}{\partial \mathbf{x}}B_k(\mathbf{x})\frac{\partial}{\partial \mathbf{x}}(u_c + \delta h_{c,k}(\delta)). \tag{9.7.13}$$

Assuming that we can take limits as $\delta \to 0$, and letting $\lim_{\delta\to 0} h_{c,k}(\delta) = h_{c,k}$, this gives that $h_{c,k}$ is the solution to the problem

$$L_c h = \frac{\partial}{\partial \mathbf{x}} B_k(\mathbf{x}) \frac{\partial}{\partial \mathbf{x}} u_c$$

$$h \in \mathcal{B}_0.$$

Thus if everything is sufficiently "nice," $X_{ijk}^{(l)}$ can be obtained by solving

$$L_{c^{(l)}} h = \frac{\partial}{\partial \mathbf{x}} B_k(\mathbf{x}) \frac{\partial}{\partial \mathbf{x}} u_{c^{(l)}} \tag{9.7.14}$$

and evaluating the solution at $\mathbf{x}(i), t_j$.

O'Sullivan (1988a) has carried out this program on a one-dimensional example.

We emphasize that this is a nonlinear ill-posed problem complicated by the fact that the degree of nonlinearity as well as the degree of ill-posedness can depend fairly strongly on the unknown solution. To see more clearly some of the issues involved, let us examine a problem sitting in Euclidean n-space that has many of the features of the system identification problem. Let $X_1, \ldots, X_N$ be $N \leq n$ matrices each of dimension $(n-M) \times n$, let B be an $M \times n$ matrix of rank M, and let $u \in E^n$, $q \in E^{n-M}$, and $b \in E^M$ be related by

$$\left(\sum_{k=1}^{N} c_k X_k \right) u = q \tag{9.7.15}$$

$$Bu = b.$$

Think of c, q, and b, respectively, as stand-ins for ρ, the forcing function, and the initial/boundary conditions.

Suppose q and b are known exactly and it is known a priori that $c_k \geq \alpha_k > 0$, $k = 1, \ldots, N$, and that this condition on the c_k's ensures that the matrix $\begin{pmatrix} \sum_k c_k X_k \\ B \end{pmatrix}$ is invertible. Suppose that one observes

$$y_i = u_i + \epsilon_i, \quad i = 1, \ldots, n$$

where u_i is the ith component of u. Letting $\Psi_{ij}(c)$ be the ijth entry of $\begin{pmatrix} \sum_k c_k X_k \\ B \end{pmatrix}^{-1}$, we may estimate c as the minimizer of

$$\sum_{i}^{n} \left(y_i - \sum_{j=1}^{n-M} \Psi_{ij}(c) q_j - \sum_{j=n-M+1}^{n} \Psi_{ij}(c) b_{j-(n-M)} \right)^2 + \lambda c' \Sigma c, \tag{9.7.16}$$

subject to $c_k \geq \alpha_k$. The ability to estimate the c's can be expected to be quite sensitive to the true values of c as well as q and b.

Returning to the original system identification problem, we now consider the case where the boundary conditions are not completely known. If (as in a one-dimensional, steady-state problem) there are only $M << n$ unknowns in the initial/boundary values, then the analogue of (9.7.16) could (in principle) be minimized with respect to c and $b = (b_1, \ldots, b_M)$.

More generally, suppose that the forcing function q and the boundary conditions $\partial u/\partial w = 0$ are known exactly, but the initial conditions $u(\mathbf{x}, 0) = u_0(\mathbf{x})$ are observed with error, that is

$$z_i = u_0(\mathbf{x}(i)) + \epsilon_i.$$

Modeling $u_0(\mathbf{x})$ as

$$u_0(\mathbf{x}) \simeq \sum_{\nu=1}^{M} b_\nu \tilde{B}_\nu(\mathbf{x}) \tag{9.7.17}$$

where the $\tilde{B}_\nu$ are appropriate basis functions (not necessarily the same as before) and letting $b = (b_1, \ldots, b_M)$, we have

$$u \simeq u(\mathbf{x}, t; c, b)$$

and we want to choose b and c, subject to appropriate constraints, to minimize

$$\frac{1}{n}\left\{\sum_{ij}(y_{ij} - u(\mathbf{x}(i), t_j; c, b))^2 + \sum_i \left(z_i - \sum_\nu b_\nu \tilde{B}_\nu(\mathbf{x}(i))\right)^2\right\}$$

$$+ \lambda_1 c'\Sigma c + \lambda_2 b'\tilde{\Sigma} b \tag{9.7.18}$$

where $b'\tilde{\Sigma}b$ is an appropriate penalty on u_0. The penalty functionals $c'\Sigma c$ and $b'\tilde{\Sigma}b$ may be quite different, since the first contains prior information about the permeability and the second about the field. This expression assumes that all the measurement errors have the same variance.

For fixed λ_1 and λ_2 this minimization can, in principle, be done as before, provided we have a means of calculating

$$z_{ij\nu} = \frac{\partial u}{\partial b_\nu}(\mathbf{x}(i), t_j; c; b). \tag{9.7.19}$$

The $z_{ij\nu}$ can be found by the same method used for the X_{ijk}. Let $u_{c,b}$ be the solution to the problem

$$L_c u_{c,b} = q, \quad \frac{\partial u_{c,b}}{\partial w} = 0, \qquad u_{c,b}(\mathbf{x}, 0) = \Sigma b_\nu \tilde{B}_\nu(\mathbf{x}). \tag{9.7.20}$$

Let $\delta_\nu = (0, \ldots, \delta, \ldots, 0)$, δ in the νth position, and let $u_{c,b+\delta_\nu}$ be the solution to

$$L_c u_{c,b+\delta_\nu} = q, \quad \frac{\partial u_{c,b+\delta_\nu}}{\partial w} = 0, \quad u_{c,b+\delta_\nu}(\mathbf{x}, 0) = \delta\tilde{B}_\nu + \left(\sum_\mu b_\mu \tilde{B}_\mu(\mathbf{x})\right) \tag{9.7.21}$$

and let

$$\tilde{h}_{c,b,\nu}(\delta) = \frac{u_{c,b+\delta_\nu} - u_{c,b}}{\delta}. \tag{9.7.22}$$

Then, subtracting (9.7.21) from (9.7.20) as before, we see that $\tilde{h}_{c,b,\nu}(\delta) = \tilde{h}_{c,b,\nu}(0)$ is the solution to the problem

$$L_c u = 0, \quad \frac{\partial u}{\partial w} = 0, u(\mathbf{x}, 0) = \tilde{B}_\nu(\mathbf{x}). \tag{9.7.23}$$

$V(\lambda_1, \lambda_2)$ can be minimized, at least in principle, to estimate good values of λ_1 and λ_2 by GCV.

CHAPTER 10

Additive and Interaction Splines

10.1 Variational problems with multiple smoothing parameters.

Let $\tilde{\mathcal{H}}$ be an r.k. space and let $\mathcal{H}$ be a (possibly proper) subspace of $\tilde{\mathcal{H}}$ of the form

$$\mathcal{H} = \mathcal{H}_0 \oplus \mathcal{H}_1$$

where $\mathcal{H}_0$ is span $\{\phi_1, \ldots, \phi_M\}$ and $\mathcal{H}_1$ is the direct sum of p orthogonal subspaces $\mathcal{H}^1, \ldots, \mathcal{H}^p$,

$$\mathcal{H}_1 = \sum_{\beta=1}^{p} \oplus \mathcal{H}^\beta. \tag{10.1.1}$$

Suppose we wish to find $f \in \mathcal{H}$ to minimize

$$\frac{1}{n}\sum_{i=1}^{n}(y_i - L_i f)^2 + \lambda \sum_{\beta=1}^{p} \theta_\beta^{-1} \|P^\beta f\|^2 \tag{10.1.2}$$

where P^β is the orthogonal projection in $\tilde{\mathcal{H}}$ onto $\mathcal{H}^\beta$ and $\theta_\beta \geq 0$. If $\theta_\beta = 0$, then the minimizer of (10.1.2) is taken to satisfy $\|P^\beta f\|^2 = 0$.

We can find the minimizer of (10.1.2) using the results of Chapter 1 by making some substitutions. Let the r.k. for $\mathcal{H}^\beta$ be $R_\beta(t,t')$. Then the r.k. for $\mathcal{H}_1$ with the squared norm $\|P_1 f\|^2_{\tilde{\mathcal{H}}} = \sum_{\beta=1}^{p} \|P^\beta f\|^2_{\tilde{\mathcal{H}}}$ is $R^1(t,t') = \sum_{\beta=1}^{p} R_\beta(t,t')$; this follows since the r.k. for a direct sum of orthogonal subspaces is the sum of the individual r.k.'s (see Aronszajn (1950)). If we change the squared norm on $\mathcal{H}_1$ from $\sum_{\beta=1}^{p} \|P^\beta f\|^2_{\tilde{\mathcal{H}}}$ to $\sum_{\beta=1}^{p} \theta_\beta^{-1} \|P^\beta f\|^2_{\tilde{\mathcal{H}}}$, then the r.k. for $\mathcal{H}^1$ changes from $\sum_{\beta=1}^{p} R_\beta(t,t')$ to $\sum_{\beta=1}^{p} \theta_\beta R_\beta(t,t')$. Using these facts and making the relevant substitutions in Chapter 1, it can be seen that the minimizer of (10.1.2) is of the form (1.3.8) with

$$\xi_i \equiv \sum_{\beta=1}^{p} P^\beta \xi_i$$

replaced everywhere by

$$\xi_i^\theta = \sum_{\beta=1}^{p} \theta_\beta P^\beta \xi_i \tag{10.1.3}$$

and that Σ in (1.3.9) is of the form

$$\Sigma = \theta_1 \Sigma_1 + \theta_2 \Sigma_2 + \ldots + \theta_p \Sigma_p, \tag{10.1.4}$$

where the ijth entry of Σ_β is

$$< P^\beta \xi_i, P^\beta \xi_j >_{\tilde{\mathcal{H}}} = L_{i(s)} \, L_{j(t)} R_\beta(s,t).$$

The minimizer of (10.1.2) is then given by

$$f_{\lambda,\theta} = \sum_{\nu=1}^{M} d_\nu \phi_\nu + \sum_{i=1}^{n} c_i \sum_{\beta=1}^{p} \theta_\beta P^\beta \xi_i \tag{10.1.5}$$

where the coefficient vectors c and d satisfy (1.3.16) and (1.3.17) with Σ given by (10.1.4). $I - A(\lambda)$ of (1.3.23) then becomes

$$I - A(\lambda, \theta) = n\lambda Q_2 (\theta_1 Q_2' \Sigma_1 Q_2 + \ldots + \theta_p Q_2' \Sigma_p Q_2 + n\lambda I)^{-1} Q_2', \tag{10.1.6}$$

where $T'Q_2 = 0_{M \times (n-M)}$ as before, so that the GCV function $V(\lambda) = V(\lambda,\theta)$ of (4.3.1) becomes

$$V(\lambda, \theta) = \frac{z'(\theta_1 \tilde{\Sigma}_1 + \ldots + \theta_p \tilde{\Sigma}_p + n\lambda I)^{-2} z}{[\mathrm{tr}(\theta_1 \tilde{\Sigma}_1 + \ldots + \theta_p \tilde{\Sigma}_p + n\lambda I)^{-1}]^2} \tag{10.1.7}$$

where

$$z = Q_2' y, \ \tilde{\Sigma}_\beta = Q_2' \Sigma_\beta Q_2.$$

The problem of choosing λ and $\theta = (\theta_1, \ldots, \theta_p)'$ by GCV is then the problem of choosing λ and θ to minimize (10.1.7), where, of course, any (λ, θ) with the same values of $\lambda_r = \lambda/\theta_r$, $r = 1, \ldots, p$ are equivalent. Numerical algorithms for doing this will be discussed in Section 11.3. For future reference we note that

$$P^\beta f_{\lambda,\theta} = \theta_\beta \sum_{i=1}^{n} c_i P^\beta \xi_i \tag{10.1.8}$$

and so

$$\|P^\beta f_{\lambda,\theta}\|^2_{\tilde{\mathcal{H}}} = \theta_\beta^2 c' \Sigma_\beta c \tag{10.1.9}$$

and

$$\sum_{i=1}^{n} (L_i P^\beta f_{\lambda,\theta})^2 = \theta_\beta^2 \|\Sigma_\beta c\|^2. \tag{10.1.10}$$

Quantities (10.1.9) and (10.1.10) can be used to decide whether the βth components are significant.

10.2 Additive and interaction smoothing splines.

The additive (smoothing) splines and their generalizations, the interaction (smoothing) splines, can be put in the framework of Section 10.1, where $\tilde{\mathcal{H}}$ is the tensor product space $\otimes^d W_m$.

The additive splines are functions of d variables, which are a sum of d functions of one variable (main effects splines)

$$f(x_1, \ldots, x_d) = f_0 + \sum_{\alpha=1}^{d} f_\alpha(x_\alpha),$$

the two factor interaction splines are of the form

$$f(x_1, \ldots, x_d) = f_0 + \sum_{\alpha=1}^{d} f_\alpha(x_\alpha) + \sum_{\alpha<\beta} f_{\alpha\beta}(x_\alpha, x_\beta),$$

and so forth, where certain side conditions on the f_α's, $f_{\alpha\beta}$'s etc., that guarantee uniqueness must hold. The additive spline models have become popular in the analysis of medical data and other contexts (see Stone (1985, 1986), Burman (1985), Friedman, Grosse, and Stuetzle, (1983), Hastie and Tibshirani (1986), Buja, Hastie, and Tibshirani (1989), and references cited therein). The interaction spline models have been discussed by Barry (1983,1986), Wahba (1986), Gu et al. (1989), Gu and Wahba (1988), and Chen (1987, 1989). These models, which in a sense generalize analysis of variance to function spaces, have strong potential for the empirical modeling of responses to economic and medical variables, given large data sets of responses with several independent variables, and represent a major advance over the usual multivariate parametric (mostly linear) models. They represent a nonparametric compromise in an attempt to overcome the "curse of dimensionality," since estimating a more general function $f(x_1, \ldots, x_d)$ will require truly large data sets for even moderate d.

To describe these splines, it will be convenient to endow $W_m[0,1]$ with a norm slightly different from the one given in Section 1.2.

Let

$$M_\nu f = \int_0^1 f^{(\nu)}(x)\,dx, \quad \nu = 0, 1, \ldots, m-1 \tag{10.2.1}$$

and note that

$$M_\nu f = f^{(\nu-1)}(1) - f^{(\nu-1)}(0), \quad \nu = 1, \ldots, m-1.$$

Let

$$\|f\|_{W_m}^2 = \sum_{\nu=0}^{m-1} (M_\nu f)^2 + \int_0^1 (f^{(m)}(u))^2 du. \tag{10.2.2}$$

Let $k_l(x) = B_l(x)/l!$, where B_l is the lth Bernoulli polynomial (see Abramowitz and Stegun (1965)); we have $M_\nu B_l = \delta_{\nu-l}$ where $\delta_i = 1$, $i = 0$, and zero otherwise. With this norm, W_m can be decomposed as the direct sum of m

orthogonal one-dimensional subspaces $\{k_l\}$, $l = 0, 1, \ldots, m-1$, where $\{k_l\}$ is the one-dimensional subspace spanned by k_l, and $\mathcal{H}_*$, which is the subspace (orthogonal to $\oplus_l \{k_l\}$) satisfying $M_\nu f = 0$, $\nu = 0, 1, \ldots, m-1$. That is,

$$W_m = \{k_0\} \oplus \{k_1\} \oplus \ldots \oplus \{k_{m-1}\} \oplus \mathcal{H}_*.$$

This construction can be found in, e.g., Craven and Wahba (1979). Letting $\otimes^d W_m$ be the tensor product of W_m with itself d times, we have

$$\overset{d}{\otimes} W_m = \overset{d}{\otimes} [\{k_0\} \oplus \ldots \oplus \{k_{m-1}\} \oplus \mathcal{H}_*]$$

and $\otimes^d W_m$ may be decomposed into the direct sum of $(m+1)^d$ fundamental subspaces, each of the form

$$[\,] \otimes [\,] \otimes \ldots \otimes [\,] \text{ (d boxes)} \tag{10.2.3}$$

where each box ([ ]) is filled with either $\{k_l\}$ for some l, or $\mathcal{H}_*$. Additive and interaction spline models are obtained by letting $\mathcal{H}_0$ and the $\mathcal{H}^\beta$'s of Section 10.1 be direct sums of various of these $(m+1)^d$ fundamental subspaces ($\mathcal{H}_0$ must, of course, be finite-dimensional). To obtain (purely) additive spline models, one retains only those subspaces of the form (10.2.3) above whose elements have a dependency on at most one variable. This means that (at most) one box is filled with an entry other than $\{k_0\} \equiv \{1\}$.

The form of the induced norms on the various subspaces can most easily be seen by an example. Suppose $d = 4$ and consider, for example, the subspace

$$[\{k_l\}] \otimes [\mathcal{H}_*] \otimes [\mathcal{H}_*] \otimes [\{k_r\}],$$

which we will assign the index $l^{**}r$. Then the squared norm of the projection of f in $\otimes^4 W_m$ onto this subspace is

$$\|P_{l^{**}r} f\|^2 = \int_0^1 \int_0^1 \left[\frac{\partial^{2m}}{\partial x_2^m \partial x_3^m} M_{l(x_1)} M_{r(x_4)} f(x_1, x_2, x_3, x_4) \right]^2 dx_2\, dx_3,$$

where $M_{k(x_\alpha)}$ means M_k applied to what follows as a function of x_α.

The reproducing kernel (r.k.) for $\{k_l\}$ is $k_l(x)k_l(x')$ and the r.k. for $\mathcal{H}_*$ (found in Craven and Wahba (1979)) is $K(x, x')$ given by

$$K(x, x') = k_m(x)k_m(x') + (-1)^{m-1} k_{2m}([x - x']) \tag{10.2.4}$$

where $[\tau]$ is the fractional part of τ.

Since the r.k. for a tensor product of two r.k. spaces is the product of the two r.k.'s (see Aronszajn (1950) for a proof), the r.k. for this subspace, call it $K_{l^{**}r}(x_1, x_2, x_3, x_4; x_1', x_2', x_3', x_4') = K_{l^{**}r}(\mathbf{x}; \mathbf{x}')$, is

$$K_{l^{**}r}(\mathbf{x}; \mathbf{x}') = k_l(x_1)k_l(x_1')K(x_2, x_2')K(x_3, x_3')k_r(x_4)k_r(x_4').$$

For more on the properties of tensor products of r.k. spaces, see Aronszajn (1950) and Weinert (1982). Tensor products of W_m were also studied by Mansfield (1972). If $L_i f = f(\mathbf{x}(i))$, where $\mathbf{x}(i)$ is the ith value of $\mathbf{x}$, then

$$(P_{l\cdots r}\xi_i)(\mathbf{x}) = K_{l\cdots r}(\mathbf{x}(i), \mathbf{x})$$

and

$$< P_{l\cdots r}\xi_i, P_{l\cdots r}\xi_j >= K_{l\cdots r}(\mathbf{x}(i), \mathbf{x}(j)).$$

In the purely additive model, $f(x_1, \ldots, x_d)$ is of the form

$$f(x_1, \ldots, x_d) = \mu + \sum_{\alpha=1}^{d} g_\alpha(x_\alpha) \tag{10.2.5}$$

where $g_\alpha \in \{k_1\} \oplus \ldots \oplus \{k_{m-1}\} \oplus \mathcal{H}_*$ and the penalty term in (10.1.2) is taken as

$$\lambda \sum_{\alpha=1}^{d} \theta_\alpha^{-1} \int_0^1 \left[\frac{\partial^m g_\alpha}{\partial x_\alpha^m}\right]^2 dx_\alpha. \tag{10.2.6}$$

To make the identifications with (10.2.3) and Section 10.1, for the purely additive spline model, $\mathcal{H}_0$ is the direct sum of the $M = 1+(m-1)d$ fundamental subspaces of the form (10.2.3) with $\{k_0\}$ in all the boxes except at most one, which contains some $\{k_l\}$ with $l > 0$. $\mathcal{H}_1 = \oplus_{\alpha=1}^{d} \mathcal{H}^\alpha$ where $\mathcal{H}^\alpha$ is of the form (10.2.3) with $\mathcal{H}_*$ in the αth box and $\{k_0\}$ in the other boxes.

If f of the form (10.2.5) is the additive spline minimizer of (10.1.2), with $L_i f = f(\mathbf{x}(i))$, then the g_α in (10.2.5) have a representation

$$g_\alpha(x_\alpha) = \sum_{\nu=1}^{m-1} d_{\nu\alpha} k_\nu(x_\alpha) + \theta_\alpha \sum_{i=1}^{n} c_i K(x_\alpha, x_\alpha(i))$$

where K is given by (10.2.4).

To discuss (two factor) interaction splines, it is convenient to consider the cases $m = 1$ and $m = 2, 3, \ldots$, separately. For $m = 1$, we have

$$\overset{d}{\otimes} W_m = \overset{d}{\otimes} [\{k_0\} \oplus \mathcal{H}_*].$$

In this case $\mathcal{H}_0$ consists of the single fundamental subspace $\otimes^d[\{k_0\}]$, there are d main effects subspaces, and there is one type of 2-factor interaction subspace, namely, one where the d boxes of (10.2.3) have $\mathcal{H}_*$ in two boxes and $\{k_0\}$ in the other $d-2$. For $m = 2, 3, \ldots$ we have two-factor interaction spaces that involve two $\{k_l\}$'s, with $l > 0$ (parametric-parametric). These may all be grouped in $\mathcal{H}_0$. Complicating matters, we may have interactions involving a $\{k_l\}$ and $\mathcal{H}_*$ (parametric-smooth) as well as two $\mathcal{H}_*$'s (smooth-smooth). For example, for $m = 2$ there are d (smooth) main effects subspaces, $d(d-1)$ fundamental subspaces with $\{k_1\} - \mathcal{H}_*$ interactions, and $d(d-1)/2$ subspaces with $\mathcal{H}_* - \mathcal{H}_*$ interactions. Similar calculations can be made for larger m and 3-factor and higher interactions. For $d = 4$, $m = 1$, we have $4 + 6 = 10$ main effects and

2-factor interaction spaces. Fitting such a model with 10 smoothing parameters has proved to be difficult but not impossible in some examples we have tried. (See the discussion in Section 11.3.) However, $m = 1$ plots tend to be visually somewhat unpleasantly locally wiggly. This is not surprising since the unknown f is not even assumed to have continuous first derivatives. However, for $d = 4$, $m = 2$, there are 4 (smooth) main effects subspaces, 12 $\{k_1\} - \mathcal{H}_*$ interaction spaces and 6 $\mathcal{H}_* - \mathcal{H}_*$ interaction spaces. We would not recommend trying to estimate 22 smoothing parameters with the present technology.

To fit additive and interaction spline models then, strategies for model simplification (selection) and numerical methods for multiple smoothing parameters are needed. Gu (1988) has suggested that the $m = 1$ case be used as a screening device. If one fits an $m = 1$ model and decides that the $\alpha\beta$th interaction is not present, then one may feel confident in eliminating both types of $\alpha\beta$ interaction in an $m = 2$ model. Similarly, if smooth main effects can be deleted in an $m = 1$ model, it can be assumed they are not present in an $m = 2$ model. (Recall that $f \in W_m \Rightarrow f \in W_{m-1}$, etc.) In Section 11.3 we discuss numerical methods for finding the GCV estimates of multiple smoothing parameters that have been used successfully in some examples with p as large as 10. If the GCV estimate $\hat{\lambda}_\beta^{-1} = \hat{\theta}_\beta/\hat{\lambda}$ is zero, then $\|P^\beta f_{\hat{\lambda},\hat{\theta}}\|^2 = 0$, and the subspace $\mathcal{H}^\beta$ can be deleted from the model. However, the probability that $\hat{\lambda}_\beta^{-1}$ is greater than zero when the true f satisfies $\|P^\beta f\|^2 = 0$ may be fairly large. (Recall the numerical results for the null model in the hypothesis tests of Section 6.3.) Thus it is desirable to have further strategies for deleting component subspaces. Possible strategies are the following. Delete $\mathcal{H}^\beta$ if, say, the contribution of this subspace to the estimated signal is small as judged by the size of

$$\sum_{i=1}^{n} (L_i P^\beta f_{\lambda,\theta})^2 = \theta_\beta^2 \|\Sigma_\beta c\|^2. \tag{10.2.7}$$

Observing that

$$\begin{pmatrix} L_1 f_{\lambda,\theta} \\ \vdots \\ L_n f_{\lambda,\theta} \end{pmatrix} = Td + \sum_{\beta=1}^{p} \theta_\beta \Sigma_\beta c$$

one could consider deleting as many of the smaller terms as possible so that

$$\|Td + \sum_{\substack{\text{terms} \\ \text{retained}}} \theta_\beta \Sigma_\beta c\|^2 \geq .95 \|Td + \sum_{\beta=1}^{p} \theta_\beta \Sigma_\beta c\|^2$$

just holds. One could also compare (10.2.7) to an estimate of $\hat{\sigma}^2$ based on the most complex reasonable model. In principle, one could generalize the GCV and GML tests of Section 6.3, to test $H_0, f \in \mathcal{H}_0 \oplus \sum_{\beta=1}^{p-1} \mathcal{H}_\beta$ versus $f \in \mathcal{H}_0 \oplus \sum_{\beta=1}^{p} \mathcal{H}_\beta$. Here, the test statistic would be the ratio of V or M minimized over the larger model to V or M minimized over the smaller model. Unfortunately, the distribution of the test statistic under H_0 will contain the nuisance parameters $\theta_1, \ldots, \theta_{p-1}$. In principle one could generate reference distributions by Monte Carlo methods, but simplifications would no doubt be necessary to make the problem tractable. This is an area of active research (Gu (1989a), Chen (1989)).

CHAPTER 11
Numerical Methods

11.1 Numerical methods that use special structure.

We have noted that a basis for an $(n-m)$-dimensional subspace of span $R_{t_1}, \ldots, R_{t_n}$ in $W_m[0,1]$ may be obtained that has compact support. This results in a band structure for certain matrices, and numerical methods that make use of this structure should be faster than those that do not. See Reinsch (1971) for a fast algorithm for computing the univariate smoothing spline when λ is given. Various authors have come upon this special structure from different points of view. Looking at the $m-1$ fold integrated Weiner process $X(t)$ of (1.5.2) it can be seen that it is an m-ple Markov process in the sense of Dolph and Woodbury (1952), Hida (1960). This means that the prediction of $X(s)$ for any $s > t$, given $X(u)$, $u \in [0,t]$ is a function of $X^{(\nu)}(t), \nu = 0, 1, \ldots, m-1$. Starting with this or similar reasoning, fast recursive formulas for the univariate polynomial smoothing spline have been obtained by various authors (see, e.g., Weinert and Kailath (1974)). Ansley and Kohn (1987) used this kind of reasoning to obtain a fast algorithm that included the computation of the GCV estimate of λ. Recently Shiau (1988) has used similar results to obtain fast algorithms for partial spline models in W_m with jumps. For some time it was an open question whether or not the special structure inherent in smoothing in W_m could be used to obtain an $O(n)$ algorithm for computing the univariate polynomial smoothing spline along with the GCV estimate of λ. Hutchinson and deHoog (1985) and O'Sullivan (1985b) appear to be the first to provide such an algorithm. A fast, accurate, user-friendly code based on the Hutchinson and deHoog paper, with some improvements, has been implemented by Woltring (1985, 1986) and may be obtained from netlib over the internet, as may O'Sullivan's code. Netlib is a robot library system run by Jack Dongarra and Eric Grosse and its internet address is netlib@research.att.com. If you write to netlib with "send index" in the body of the message, the robot mailserver will return instructions for using the system. O'Sullivan's and Woltring's code may be obtained this way. Code for generating B-splines based on deBoor's book may also be obtained this way. Earlier, Utreras (1983) provided a cheap way of evaluating trace $A(\lambda)$ in the equally spaced data case in W_m, based on good approximate values for the relevant eigenvalues. This method is implemented in Version 8 et seq of IMSL (1986).

11.2 Methods for unstructured problems.

Basic references for numerical linear algebra and optimization methods are Golub and Van Loan (1983) and Gill, Murray, and Wright (1981), respectively. The LINPACK manual (Dongarra et al. (1979)) is a good place to read about the singular value decomposition, the QR decomposition, Cholesky decomposition, Householder transformations, and other useful tools for computing the estimates in this book. See also the EISPACK manual (Smith et al. (1976)). Generally, the most computer-intensive part of the calculation of cross-validated estimates discussed in this book is going to be the calculation of trace $I - A(\lambda)$ in the denominator of the GCV function. The optimization of that calculation will usually dictate the manner in which the remaining calculations are done.

Considering the case where N basis functions are used, as in Secton 7.1, let

$$A(\lambda) = X(X'X + n\lambda\Sigma)^{-1}X'$$

$$= \tilde{X}(\tilde{X}'\tilde{X} + n\lambda I)^{-1}\tilde{X}'$$

where $\tilde{X} = X\Sigma^{-1/2}$, where $\Sigma^{-1/2}$ is any square root of Σ^{-1}. For small-to-moderate problems $\Sigma^{-1/2}$ may be obtained numerically as L^{-1} where LL' is the Cholesky decomposition of Σ, L being lower triangular, and hence numerically easy to invert (if it is not too ill-conditioned). Then

$$\mathrm{tr}\,(I - A(\lambda)) = n - \sum_{\nu=1}^{N} \frac{s_\nu^2}{s_\nu^2 + n\lambda}$$

where the s_ν's are the singular values of $\tilde{X}$. The singular value decomposition (SVD) in LINPACK (Dongarra et al. (1979)) can be used to compute the s_ν's. Elden (1984) proposed a method for computing trace $A(\lambda)$ based on a bidiagonalization of $\tilde{X}$ that is much faster than using the singular value decomposition of $\tilde{X}$ (see also Elden (1977)). Bates and Wahba (1982) proposed a method for computing trace $A(\lambda)$ based on truncating a pivoted QR decomposition of $\tilde{X}$, $\tilde{X}$ =QR. (Recall that Q is $n \times n$ orthogonal and R is $n \times N$ upper triangular; see Dongarra et al., Chap. 9 for the pivoted QR decomposition.) The pivoted QR decomposition permutes the columns of $\tilde{X}$ so that R has the property that its entries r_{ij} satisfy

$$r_{kk}^2 \geq \sum_{i=k}^{j} r_{ij}^2.$$

If R is replaced by R_{trunc}, which is obtained by setting r_{ij} to zero for $i = k+1, \ldots, N$, then the Weilandt–Hoffman theorem (Golub and Van Loan (1983, p.270) says that

$$\mathrm{trace}\,(\tilde{X} - \tilde{X}_{\mathrm{trunc}})(\tilde{X} - \tilde{X}_{\mathrm{trunc}})' = \sum_{i=k+1}^{N} \sum_{j=i}^{N} r_{ij}^2 = \tau, \text{ say.}$$

Thus k can be chosen so that the tolerance τ is less than a prespecified small amount. This method can be expected to be useful when n and N are large and $\tilde{X}$ has a large number of eigenvalues near zero. In these kinds of cases it has sometimes been found that the LINPACK SVD will converge quickly on R_{trunc} when it converges slowly or not at all on $\tilde{X}$. It is implemented in GCVPACK (Bates et al. (1987)) and the code is available through netlib. GCVPACK also has a code for partial spline models where the smooth part is a thin-plate spline. Hutchinson and Bischof (1983) and Hutchinson (1984, 1985) have developed transportable code for thin-plate splines using the thin-plate basis functions of Wahba (1980b) described in (7.1.4) and (7.1.5). Recently, Girard (1987a,b) has proposed an ingenious method for estimating trace $A(\lambda)$ when n is very large, based on the fact that, if $\epsilon \sim \mathcal{N}(0, I)$, then $E\epsilon' A(\lambda)\epsilon = \text{trace}A(\lambda)$. A random vector ϵ is generated and $A(\lambda)\epsilon$ obtained in $O(n)$ operations by solving a linear system. A formula for the standard deviation of $\epsilon' A(\lambda)\epsilon$ is given, and if this standard deviation is too large, then the average of k replications of this estimate can be taken; the standard deviation will go down as $1/\sqrt{k}$. Hutchinson (1989) has studied this approach with the ϵ_i plus or minus one with probability $\frac{1}{2}$.

Gu et al. (1989) have considered the general unstructured case when V is defined by (4.3.1) with A as in (1.3.23),

$$V(\lambda) = \frac{1}{n} z'(\tilde{\Sigma} + n\lambda I)^{-2} z \Big/ \left(\frac{1}{n} \operatorname{tr} (\tilde{\Sigma} + n\lambda I)^{-1} \right)^2 \tag{11.2.1}$$

where $\tilde{\Sigma} = Q_2' \Sigma Q_2$ and $z = Q_2' y$. An outline of the major steps goes as follows.

(1) Tridiagonalize $\tilde{\Sigma}$ as

$$U' \tilde{\Sigma} U = \Delta$$

where U is orthogonal and Δ is tridiagonal. This can be done by successively applying the Householder transformation (see Dongarra et al. (1979)). A distributed truncation method is provided in Gu et al. (1988) for speeding up this step. Letting $x = Uz$, then

$$V(\lambda) = \frac{1}{n} x'(n\lambda I + \Delta)^{-2} x \Big/ \left(\frac{1}{n} \operatorname{tr} (n\lambda I + \Delta)^{-1} \right)^2 .$$

(2) Compute the Cholesky decomposition $(n\lambda I + \Delta) = C'C$, where

$$C = \begin{bmatrix} a_1 & b_1 & & & \\ & a_2 & b_2 & & \\ & & \ddots & \ddots & \\ & & & a_{n-M-1} & b_{n-M-1} \\ & & & & a_{n-M} \end{bmatrix}$$

is upper diagonal.

(3) Calculate $\text{tr}(C^{-1} C^{-1'})$ using a trick due to Elden (1984). Letting the ith row of C^{-1} be $\mathbf{c}_i'$, then we have $\text{tr}(C^{-1} C^{-1'}) = \sum_{i=1}^{n-M} \|\mathbf{c}_i\|^2$.

From

$$C^{-1'}C' = (\mathbf{c}_1, \ldots \mathbf{c}_{n-M}) \begin{bmatrix} a_1 & & & & \\ b_1 & a_2 & & & \\ & b_2 & & & \\ & & \ddots & & \\ & & \ddots & a_{n-M-1} & \\ & & & b_{n-M-1} & a_{n-M} \end{bmatrix} = I$$

we have

$$\begin{aligned} a_{n-M} \quad \mathbf{c}_{n-M} &= \mathbf{e}_{n-M} \\ a_i \quad \mathbf{c}_i &= \mathbf{e}_i - b_i \mathbf{c}_{i+1}, \quad i = n-M-1, \ldots, 1 \end{aligned}$$

where the $\mathbf{e}_i$'s are unit vectors. Because $C^{-1'}$ is lower triangular, $\mathbf{c}_{i+1}$ is orthogonal to $\mathbf{e}_i$, giving the recursive formula

$$\begin{aligned} \|\mathbf{c}_{n-M}\|^2 &= a_{n-M}^{-2}, \\ \|\mathbf{c}_i\|^2 &= (1 + b_i^2 \|\mathbf{c}_{i+1}\|^2) a_i^{-2}, \; i = n-M-1, \ldots, 1. \end{aligned}$$

11.3 Methods for multiple smoothing parameters, with application to additive and interaction splines

The algorithm of Gu et al. (1989) has been used as a building block by Gu and Wahba (1988) in an algorithm for minimizing $V(\lambda, \theta)$ of (10.1.7), thus allowing the calculation of additive and interaction splines with multiple smoothing parameters chosen by GCV. The software appears in Gu (1989b) and can be obtained from netlib. Here recall that $V(\lambda, \theta)$ is given by (11.2.1) with $\tilde{\Sigma}$ replaced by $\theta_1 \tilde{\Sigma}_1 + \ldots + \theta_p \tilde{\Sigma}_p$,

$$V(\lambda, \theta) = \frac{\frac{1}{n} z'(\theta_1 \tilde{\Sigma}_1 + \ldots + \theta_p \tilde{\Sigma}_p + n\lambda I)^{-2} z}{\left(\frac{1}{n} \operatorname{tr} (\theta_1 \tilde{\Sigma}_1 + \ldots + \theta_p \tilde{\Sigma}_p + n\lambda I)^{-1}\right)^2}. \tag{11.3.1}$$

As noted previously, all sets (θ, λ) with $\lambda_\beta = \lambda/\theta_\beta$ are equivalent. However a minimization of (11.3.1) in λ is "cheap" compared to a minimization in components of θ. We briefly describe the algorithm. The algorithm works iteratively by first fixing θ and minimizing $V(\lambda|\theta)$ by the algorithm in Gu et al. (1988). Then, for fixed λ, the gradient and Hessian of $V(\theta|\lambda)$ are evaluated with respect to the variables $\rho_\beta = \log \theta_\beta$, and the ρ_β are updated by a modified Newton method. In this sort of optimization it is important to get good (relative) starting values for the θ's. The default starting values in Gu and Wahba (1988) are obtained as follows. The initial θ_β's are taken as

$$\theta_\beta^{(0)} = (\operatorname{tr} \tilde{\Sigma}_\beta)^{-1}.$$

Then $V(\lambda|\theta)$ is minimized. (By this we mean $V(\lambda, \theta)$ is considered as a function of λ with θ given.) This results in a trial $f_{\lambda,\theta}$, with $\|P^\beta f_{\lambda,\theta}\|^2 = (\theta_\beta^{(0)})^2 c' \tilde{\Sigma}_\beta c$, from (10.1.9). New starting values $\theta_\beta^{(1)}$ of the θ_β's are taken as

$$\theta_\beta^{(1)} = (\theta_\beta^{(0)})^2 c' \tilde{\Sigma}_\beta c,$$

where c is associated with the trial $f_{\lambda,\theta}$. $V(\lambda|\theta^{(1)})$ is then minimized with respect to λ via the algorithm of Gu et al. (1989). $\theta_\beta^{(2)}$ and successive θ_β's are obtained by the modified Newton update, alternating minimizations with respect to λ. The algorithm has been observed to converge rapidly in a number of examples.

Here convergence is defined in terms of minimizing $V(\theta, \lambda)$, and is declared to have occurred if the gradient is small and V no longer decreases. In examples this has also been seen to drive the predictive mean-square error $T(\lambda, \theta)$ to a minimum. This does not *necessarily* mean that a particular $\lambda_\beta = \lambda/\theta_\beta$ has converged, since if the predictive mean-square error $T(\lambda, \theta)$ is insensitive to certain variations in the λ_β, then so will be $V(\lambda, \theta)$, and these cannot be sorted out by minimizing $V(\lambda, \theta)$. For predictive mean-square error purposes, one would presumably be indifferent to them. Loosely speaking, λ_α will be more or less distinguishable from λ_β, according to whether $\mathrm{Tr}\, \tilde{\Sigma}_\alpha \tilde{\Sigma}_\beta$ is small or large.

In Figures 11.1 and 11.2 we present the results of a Monte Carlo example of an additive model with $d = 4$, $m = 2$, $n = 100$. The data were generated by

$$y_i = f(\mathbf{x}(i)) + \epsilon_i, \quad i = 1, \ldots, n$$

with $\epsilon_i \sim \mathcal{N}(0, \sigma^2)$ with $\sigma = 1$, and $\mathbf{x} = (x_1, x_2, x_3, x_4)$ with

$$\begin{aligned} f(x_1, x_2, x_3, x_4) &= 10 \sin(\pi x_2) + \exp(3x_3) \\ &\quad + 10^6 x_4^{11}(1 - x_4)^6 + 10^4 x_4^3 (1 - x_4)^{10} \\ &= f_2(x_2) + f_3(x_3) + f_4(x_4), \text{ say.} \end{aligned}$$

Thus the truth is additive with no parametric or main effects component for x_1. The $\mathbf{x}(i)$ were random uniform variables on the unit 4-cube. Figure 11.1 gives a scatter plot of the $x_\alpha(i)$, $i = 1, 2, 3, 4$ and y_i. The dashed lines in Figure 11.2 are the $f_\alpha(x_\alpha)$, with $f_1(x_1) \equiv 0$, and the solid lines are the estimates. The main effects are uniquely determined by the fact that their average values are zero. They have been shifted in the plots to match the means of the f_α's. (The constant component of the model was estimated to better than the visual resolution of the picture.)

Table 11.1 gives the value of $V(\lambda, \theta)$ and $T(\lambda, \theta)$ after each iteration cycle (after the λ-step). It can be seen that convergence was quite rapid and T appears to decrease along with V.

11.4 Applications to components of variance, a problem from meteorology

A similar algorithm for minimizing the GML function

$$M(\theta, \lambda) = \frac{z'(\theta_1 \tilde{\Sigma}_1 + \cdots + \theta_p \tilde{\Sigma}_p + n\lambda I)^{-1} z}{[\det (\theta_1 \tilde{\Sigma}_1 + \cdots + \theta_p \tilde{\Sigma}_p + n\lambda I)^{-1}]^{1/n-M}} \tag{11.4.1}$$

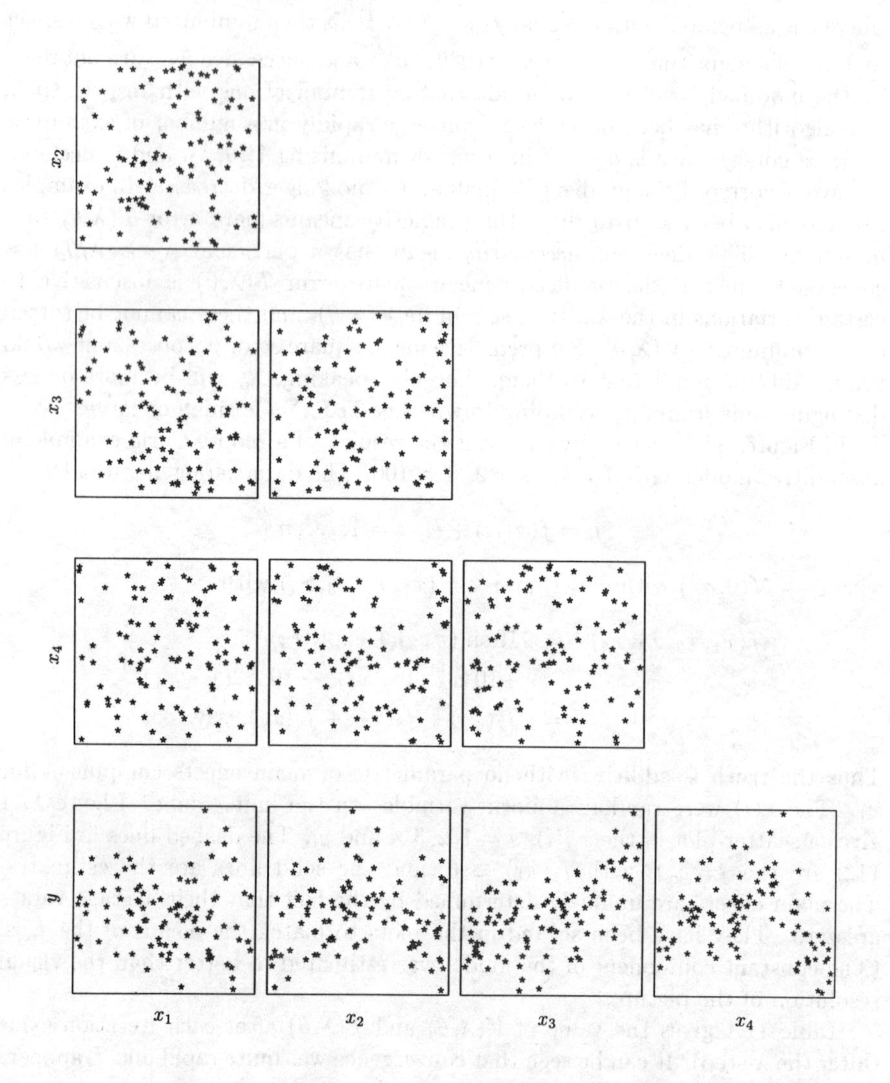

FIG. 11.1. *Scatter plot matrix for the additive model.*

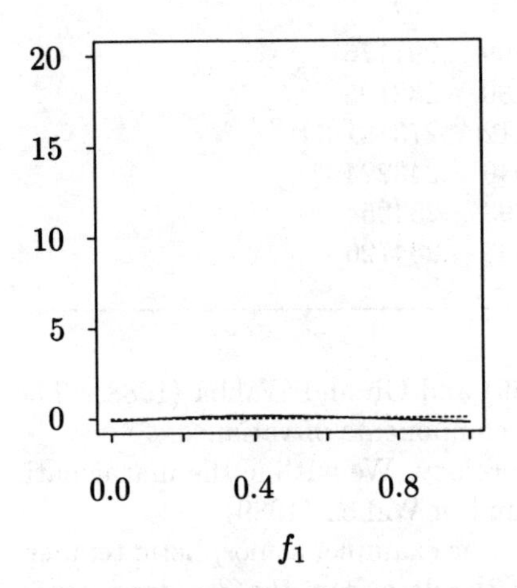

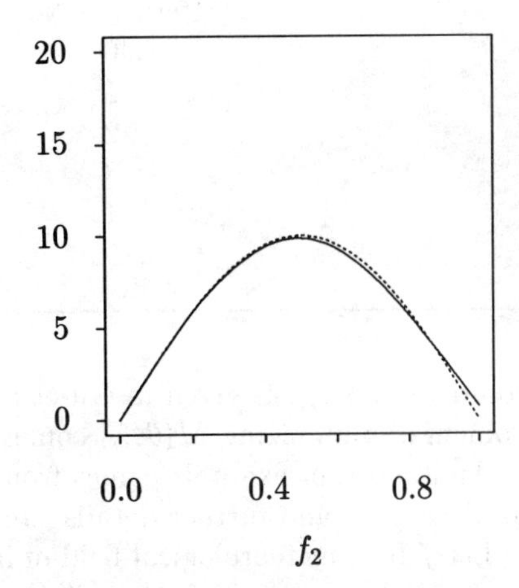

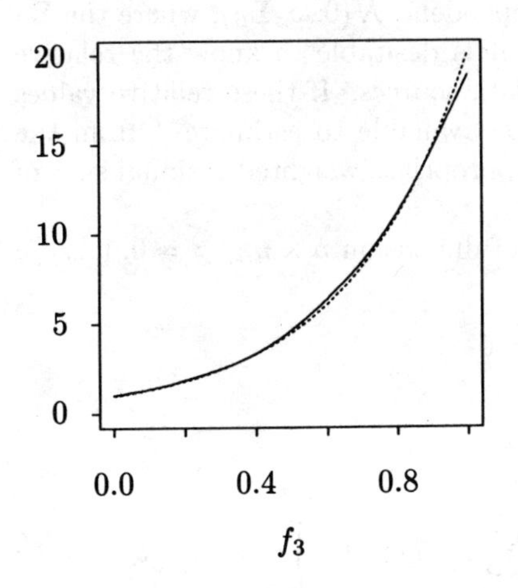

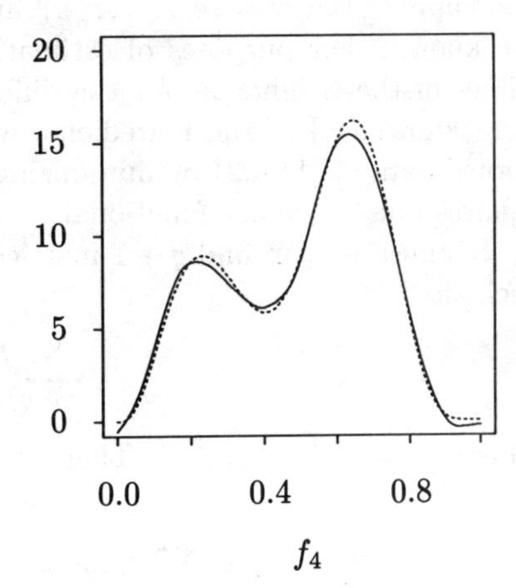

FIG. 11.2. *Estimates (solid) and "truth" (dashed) for the additive model.*

TABLE 11.1
V and T, iterated values.

Iteration No.	V	T
0	1.50409	.291176
1	1.50180	.232195
2	1.45412	.273181
3	1.41040	.243224
4	1.40893	.234954
5	1.40893	.234726

(compare (4.8.4)), is given in Gu et al. (1989) and Gu and Wahba (1988). The problem of minimizing $M(\theta, \lambda)$ comes up in components of variance.

An important example comes from meteorology. We outline the mathematical ideas here, and further details can be found in Wahba (1989).

Let f be a meteorological field of interest (for example, atmospheric temperature, wind, geopotential, humidity), and let the data from the βth data source (sensor or forecast) be

$$y_{i\beta} = L_i^\beta f + \epsilon_i^\beta \ , \ i = 1, \ldots, n_\beta, \ \beta = 0, 1, \ldots, q, \tag{11.4.2}$$

and suppose the $\epsilon^\beta = (\epsilon_1^\beta, \ldots, \epsilon_{n\beta}^\beta)'$ are independent, $\mathcal{N}(0, \omega_\beta \Sigma_\beta)$, where the Σ_β are known. For purposes of estimating f, it is desirable to know the relative values of the weights ω_β for the different data sources. If these relative values were known, all of the ingredients would be available to estimate f from the pooled data of (11.4.2) by minimizing the appropriate weighted residual sum of squares plus a penalty functional.

Suppose we can find $q+1$ matrices B_β of dimension $n \times n_\beta$, $\beta = 0, 1, \ldots, q$ such that

$$\sum_{\beta=0}^{q} B_\beta L^\beta = 0,$$

where $L^\beta = (L_1^\beta, \ldots, L_{n_\beta}^\beta)'$. Then

$$u \equiv \sum_{\beta=0}^{q} B_\beta y^\beta \sim \mathcal{N}\left(0, \sum_{\beta=0}^{q} \omega_\beta B_\beta \Sigma_\beta B_\beta'\right)$$

where $y^\beta = (y_1^\beta, \ldots, y_{n_\beta}^\beta)'$. Suppose that $B_0 \Sigma_0 B_0'$ is well-conditioned. Then taking the eigenvalue eigenvector decomposition $B_0 \Sigma_0 B_0' = UDU'$, and letting

$$z = D^{-1/2} U' u,$$

we have

$$z \sim \mathcal{N}(0, \omega_1 \tilde{\Sigma}_1 + \cdots + \omega_q \tilde{\Sigma}_q + \omega_0 I),$$

where $\tilde{\Sigma}_\beta = D^{-1/2}U'\Sigma_\beta U D^{-1/2}$. Letting $\omega_0 = \sigma^2 n\lambda$ and $\omega_\beta = \sigma^2\theta_\beta$, one can minimize the negative log-likelihood for z with respect to σ^2 explicitly. Substituting in the resulting $\hat{\sigma}^2$, one is left with an expression of the form of (11.4.1) to minimize. In order to get good estimates of $\lambda_\beta = \lambda/\theta_\beta$, it is necessary that the correlation structure of the different data sets $B_\beta y^\beta$ be sufficiently different. Some examples where this is likely to happen in practice are given in Lonnberg and Hollingsworth (1986). For some other examples of the use of multiple smoothing parameters in meteorological applications, see Wahba (1982d), Hoffman (1984, 1985), and Legler, Navon, and O'Brien (1989). A. Hollingsworth (1989) informs us that development of new data analysis for estimating initial conditions for use in the European Center for Medium-Range Weather Forecasts is following along lines suggested by Wahba (1982d).

CHAPTER 12

Special Topics

12.1 The notion of "high frequency" in different spaces.

Let $\Psi_\nu(t) = \sqrt{2}\cos 2\pi\nu t$, $\nu = 1, 2, \ldots$ and let $\mathcal{H}$ be the collection of all functions of d variables with a representation

$$f(x_1, \ldots, x_d) = \sum_{\nu_1=1}^{\infty} \cdots \sum_{\nu_d=1}^{\infty} f_{\nu_1,\ldots,\nu_d} \Psi_{\nu_1}(x_1) \ldots \Psi_{\nu_d}(x_d) \tag{12.1.1}$$

such that

$$\sum_{\nu_1,\ldots,\nu_d=1}^{\infty} (2\pi\nu_1)^{2m} \ldots (2\pi\nu_d)^{2m} f^2_{\nu_1,\ldots,\nu_d} < \infty. \tag{12.1.2}$$

We see that the left-hand side of (12.1.2) is then equal to

$$\int_0^1 \cdots \int_0^1 \left(\frac{\partial^{dm}}{\partial x_1^m \ldots \partial x_d^m} f \right)^2 dx_1 \ldots dx_d, \tag{12.1.3}$$

which we may take as the squared norm in $\mathcal{H}$. (We could let the Ψ_ν's be sines as well as cosines and get the same result; to avoid cumbersome notation we will not do this.)

Alternatively, consider the collection of functions of the form (12.1.1) for which

$$\sum_{\nu_1,\ldots,\nu_d=1}^{\infty} [(2\pi\nu_1)^2 + \ldots + (2\pi\nu_d)^2]^m f^2_{\nu_1,\ldots,\nu_d} < \infty. \tag{12.1.4}$$

It is not hard to see that (12.1.4) is equal to

$$\sum_{\mu_1+\ldots+\mu_d=m} \frac{m!}{\mu_1! \ldots \mu_d!} \int_0^1 \cdots \int_0^1 \left(\frac{\partial^m f}{\partial x_1^{\mu_1} \cdots \partial x_d^{\mu_d}} \right)^2 dx_1 \ldots dx_d. \tag{12.1.5}$$

The Hilbert spaces with squared norms (12.1.2) and (12.1.4) are quite different. The eigenvalues of the r.k. for the first space are

$$\lambda_{\nu_1,\ldots,\nu_d} = [(2\pi\nu_1)(2\pi\nu_2) \ldots (2\pi\nu_d)]^{-2m},$$

and if these are lined up in size place, it can be shown that the nth largest is of the order of $\left[((\log n)^{d-1}/n)\right]^{2m}$. To see this, observe that the number of lattice points $(\nu_1, \ldots, \nu_d)$ on the lattice of d tuples of positive integers that satisfies

$$\prod_{\alpha=1}^{d} \nu_\alpha \leq k$$

is, to first order, given by $(1/(d-1)!)k(\log k)^{d-1}(1+o(1))$. This is obtained from the volume approximation given by

$$\int_1^k \cdots \int_1^{k/x_3 \ldots x_{d-1}} \int_1^{k/x_2 \ldots x_{d-1}} \frac{k}{x_1 \ldots x_{d-1}} dx_1 \ldots dx_{d-1} = \frac{k}{(d-1)!}(\log k)^{d-1}.$$

(For a sharper approximation, see Ivic (1985, Chap. 3).) Letting Λ_n be the nth largest eigenvalue, we obtain

$$\Lambda_{[k(\log k)^{d-1}/(d-1)!]} \simeq \frac{1}{(2\pi)^d k^{2m}}.$$

Setting

$$n = [k(\log k)^{d-1}/(d-1)!]$$

gives

$$1/(2\pi)^d k^{2m} = O\left[(\log n)^{d-1}/n\right]^{2m} = \Lambda_n.$$

Similarly, for the space of (12.1.4) the eigenvalues of the r.k. are

$$\lambda_{\nu_1, \ldots, \nu_d} = \left(\sum_{\alpha=1}^{d} (2\pi\nu_\alpha)^2 \right)^{-m},$$

and we use the volume inside a sphere to estimate the number of lattice points for which $\sum_{\alpha=1}^{d} \nu_\alpha^2 \leq k$. The result is the nth largest eigenvalue is of the order of $n^{-2m/d}$.

We have noted that the rate of decay of the eigenvalues of the r.k. plays a role in convergence rates of estimates with noisy data.

Below is a handy theorem giving lower bounds on the r.k. norm of the error, when the data are exact and the estimate is an orthogonal projection.

THEOREM 12.1.1 (Micchelli and Wahba (1981)). *Let $\mathcal{H}_R$ be an* r.k.h.s. *with* r.k. *R and eigenvalues and eigenfunctions (λ_ν, Φ_ν), $\nu = 1, 2, \ldots$. Let V_n be any n-dimensional subspace in $\mathcal{H}_R$ and let $P_{V_n} g$ be the orthogonal projection of g in $\mathcal{H}_R$ onto V_n. Then, for any $p > 1$, there exists $g \in \mathcal{H}_R$ with*

$$g(t) = \int_T R^{p/2}(t, u)\rho(u)\, du \tag{12.1.6}$$

with $\rho \in \mathcal{L}_2[T]$, $\int_T \rho^2(u)\, du = 1$, such that

$$\|g - P_{V_n} g\|^2 \geq \lambda_{n+1}^{p-1}. \tag{12.1.7}$$

To see why this theorem is true, let

$$\mathcal{C}_p = \left\{ g : g(t) = \int_T R^{p/2}(t,u)\rho(u)\,du \right\}$$

with $\rho \in \mathcal{L}_2$ and

$$\int \rho^2(s)\,ds = 1.$$

Then

$$g(t) = \int \sum_{\nu=1}^{\infty} \lambda_\nu^{p/2} \Phi_\nu(t)\Phi_\nu(u)\rho(u)\,du,$$

that is,

$$g = \sum_{\nu=1}^{\infty} \lambda_\nu^{p/2} \rho_\nu \Phi_\nu$$

where

$$\rho_\nu = \int \Phi_\nu(u)\rho(u)\,du$$

and

$$\Sigma \rho_\nu^2 \le 1.$$

Consider

$$\inf_{V_n} \sup_{g \in \mathcal{C}_p} \|g - P_{V_n} g\|^2.$$

We have the following game. You choose V_n to minimize $\|g - P_{V_n} g\|$ and nature chooses g to maximize it. The optimal strategy is to choose $V_n =$ span $(\Phi_1, \ldots, \Phi_n)$. Then nature chooses ρ_ν all zero except $\rho_{n+1} = 1$, then $g = \lambda_{n+1}^{p/2} \Phi_{n+1}$ and

$$\|g - P_{V_n} g\|^2 = \|\lambda_{n+1}^{p/2} \Phi_{n+1}\|^2 = \frac{\lambda_{n+1}^p}{\lambda_{n+1}} = \lambda_{n+1}^{p-1}.$$

There are examples (with $\mathcal{T} = [0,1]$) for which $V_n =$ span $\{R_{t_1}, \ldots, R_{t_n}\}$ is also an optimal strategy. The $t_1, \ldots, t_n$ are the n zeros of the $(n+1)$st eigenfunction (see Melkman and Micchelli (1978)).

This is, of course, a theorem giving lower bounds on interpolation error. To see this, let V_n be spanned by $\eta_1, \ldots, \eta_n$ and let $g_i = < g, \eta_i >$, $i = 1, \ldots, n$; then $P_{V_n} g$ is that element $\hat{g}$ in $\mathcal{H}_R$ minimizing $\|g\|$ subject to $< \hat{g}, \eta_i > = g_i$, $i = 1, \ldots, n$. Note that any element in $\mathcal{H}_R$ has a representation as a multiple of an element in $\mathcal{C}_p$ for $p = 1$ (to which this theorem does not apply). As an application, consider W_m, with the r.k. of Chapter 10. Then $\mathcal{C}_2$ is the collection of functions that satisfy $f^{(2m)} \in \mathcal{L}_2$ and certain boundary conditions. Since the eigenvalues of the r.k. decay at the rate ν^{-2m}, this theorem says that if one interpolates to f at the points t_i, $i = 1, \ldots, n$, then $1/n^{2m}$ is a lower bound on the best achievable convergence rate of $\|g - P_{V_n} g\|^2$ for $g \in \mathcal{C}_2$.

Note that if the loss were measured as $\|g - P_{V_n} g\|^2_{\mathcal{L}_2}$, where the subscript indicates the $\mathcal{L}_2$ norm rather than the r. k. norm, the game would be the same, with

$$\inf_{V_n} \sup_{g \in \mathcal{C}_p} \|g - P_{V_n} g\|^2_{\mathcal{L}_2} = \lambda^p_{n+1}.$$

12.2 Optimal quadrature and experimental design.

Let

$$f(t) = bg(t) + X(t), \quad t \in \mathcal{T}$$

where X is a zero-mean Gaussian stochastic process with $EX(s)X(t) = R(s,t)$. One will observe $f(t)$ for $t = t_1, \ldots, t_n$, g is known, and it is desired to estimate b. The Gauss–Markov estimate of b is

$$\hat{b} = \frac{f'\Sigma^{-1} g}{g'\Sigma^{-1} g}$$

where $f = (f(t_1) \ldots f(t_n))'$, $g = (g(t_1), \ldots, g(t_n))'$ and Σ is the $n \times n$ matrix with ijth entry $R(t_i, t_j)$. The variance of $\hat{b}$ is $(g'\Sigma^{-1} g)^{-1} = \|P_{V_n} g\|^{-2}$, where P_{V_n} is the orthogonal projection onto $R_{t_1}, \ldots, R_{t_n}$. Letting V_n be span $R_{t_1}, \ldots, R_{t_n}$, we have that this experimental design problem then is equivalent to: Choose $t_1, \ldots, t_n$ to minimize

$$\|g - P_{V_n} g\|^2.$$

Lower bounds on $\|g - P_{V_n} g\|^2$ follow from (12.1.7) in the case $g \in \mathcal{H}_{R^p}$ for $p > 1$. This problem was posed and studied by Sacks and Ylvisaker (1969), and studied by a number of authors (see Wahba (1971) and references there, also Wahba (1976, 1978c) and Athavale and Wahba (1979)). Let

$$Lh = \int \rho(u) h(u)\, du$$

and suppose one wishes to estimate Lh, given data $h(t_i) = < R_{t_i}, h >$, $i = 1, \ldots, n$. Let V_n be span $R_{t_1}, \ldots, R_{t_n}$. Let $P_{V_n} h$ be the minimal norm interpolant to this data; then

$$\widehat{Lh} = \int \rho(u)\ (P_{V_n} h)(u)\, du$$

gives a quadrature formula, that is, a formula of the form

$$\widehat{Lh} = \Sigma w_i h(t_i).$$

Letting h be the representer for L,

$$h(s) = \int R(s,u) \rho(u)\, du, \tag{12.2.1}$$

we have

$$\widehat{Lh} = < h, P_{V_n} h > = < P_{V_n} g, h >$$

and

$$|Lh - \widehat{Lh}| = | < h - P_{V_n} g, h > | \le \|h - P_{V_n} g\| \|h\|.$$

The optimal quadrature problem then becomes the problem of choosing $t_1, \ldots, t_n$ to minimize $\|g - P_{V_n} g\|^2$. Note that g of (12.2.1) is in $\mathcal{H}_{R^p}$ with $p = 2$.

The major $\mathcal{T} = [0,1]$ results are loosely described as follows. (For technical details, see the references.) Let the r.k. $R(s,t)$ be a Green's function for a $2m$th order linear differential operator (as in Section 1.2, for example), or equivalent to such an R. Let the characteristic discontinuity of R be

$$\lim_{s \downarrow t} \frac{\partial^{2m-1}}{\partial s^{2m-1}} R(s,t) - \lim_{s \uparrow t} \frac{\partial^{2m-1}}{\partial s^{2m-1}} R(s,t) = (-1)^m \alpha(t)$$

for some $\alpha(t) > 0$. Let $g(s) = \int_0^1 R(s,t)\rho(t)\,dt$ where ρ is strictly positive and has a bounded first derivative on $[0,1]$. Then an asymptotically optimal design $R_{t_1}, \ldots, R_{t_n}$ for minimizing $\|g - P_{V_n} g\|^2$ is given by $t_1, \ldots, t_n$ satisfying

$$\int_0^{t_i} [\rho^2(u)\alpha(u)]^{1/(2m+1)} = \frac{i}{n} \int_0^1 [\rho^2(u)\alpha(u)]^{1/(2m+1)}\,du, \quad i = 1, \ldots, n. \quad (12.2.2)$$

These results have been used in a sequential procedure that involves starting with a trial design, estimating ρ, and then using (12.2.2) to obtain an additional set of design points, etc., (see Athavale and Wahba (1979)).

Very little is known of optimal designs for $\mathcal{T}$ other than $[0,1]$. Optimal designs in the tensor product space associated with (12.1.2) can be expected to be different from those for the (thin-plate) space associated with (12.1.4), because the eigenfunctions associated with the largest eigenvalues are different. Some very curious examples for tensor product spaces are given in Wahba (1978c). The designs given there are for evaluation functionals and their span approximates the span of the eigenfunctions with large eigenvalues. These designs are known as blending function designs (see Delvos and Posdorf (1977)). Some recent related work can be found in Donoho and Johnstone (1989).

The noisy data case is of some importance but very little is known. Here let

$$y_i = < \eta_i, f > + \epsilon_i, \quad i = 1, \ldots, n$$

as before with $\epsilon = (\epsilon_1, \ldots, \epsilon_n) \sim \mathcal{N}(0, \sigma^2 I)$ and let f_λ be the minimizer (for simplicity) of

$$\sum_{i=1}^n (y_i - < \eta_i, f >)^2 + \lambda \|f\|^2.$$

The problem is to choose $\eta_1, \ldots, \eta_n$ so that some loss function depending on $f - f_\lambda$ is small. One may require that $\eta_i = R_{t_i}$ for some $t_1, \ldots, t_n$, or one may have more freedom to choose the η_i's. In this latter case where the $\|\eta_i\|$ must be bounded to make the problem nontrivial; we set $\|\eta_i\| = 1$. Plaskota (1989), has recently shown that if f is a Gaussian stochastic process with $Ef(s)f(t) = R(s,t)$, then to minimize expected squared $\mathcal{L}_2$-norm of the error (with the expectation taken over f as well as the ϵ_i's), the optimal η_i's are in the span of a proper subset of the first n eigenfunctions Φ_ν, with replications.

Suppose the design is $\eta_1, \ldots, \eta_N$ with η_i replicated n_i times, $\sum_{i=1}^N n_i = n$. Then the information available is assumed to be equivalent to

$$\tilde{y}_i = < \eta_i, f > + \tilde{\epsilon}_i, \qquad i = 1, \ldots, N,$$

where $E\tilde{\epsilon}_i^2 = \sigma^2/n_i$. Here $\tilde{y}_i$ is the average of the n_i observations involving η_i, and f is estimated as the minimizer of

$$\sum_{i=1}^N n_i(\tilde{y}_i - < \eta_i, f >)^2 + n\lambda \|f\|^2. \tag{12.2.3}$$

Let us see what happens if it is assumed that the η_i's are in span $\{\Phi_1, \ldots, \Phi_n\}$. Let $\eta_\nu = \sqrt{\lambda_\nu}\Phi_\nu$: this ensures that $\|\eta_\nu\|_R = 1$. Then $< \eta_\nu, f > = f_\nu/\sqrt{\lambda_\nu}$, where $f_\nu = \int f(t)\Phi_\nu(t)dt$. After some calculations one obtains $f_\lambda = \sum_{\nu=1}^N \hat{f}_\nu \Phi_\nu$, where $\hat{f}_\nu = \sqrt{\lambda_\nu}(n_\nu/(n_\nu + n\lambda))\tilde{y}_\nu$. Further calculation then gives the expected squared $\mathcal{L}_2$-norm of the error as

$$E\|f - f_\lambda\|_{\mathcal{L}_2}^2 = \sum_{\nu=1}^N \left(\frac{n\lambda}{n_\nu + n\lambda}\right)^2 f_\nu^2 + \sigma^2 \sum_{\nu=1}^N \lambda_\nu \frac{n_\nu}{(n_\nu + n\lambda)^2} + \sum_{i=N+1}^\infty f_\nu^2. \tag{12.2.4}$$

Plaskota's assumption that $Ef(s)f(t) = R(s,t)$ entails that $Ef_\nu^2 = \lambda_\nu$ and the optimum $n\lambda$ averaged over sample functions is σ^2. Making these substitutions in (12.2.4), we obtain

$$\begin{aligned} E_f E\|f - f_\lambda\|_{\mathcal{L}_2}^2 &= \sum_{\nu=1}^N \left(\frac{\sigma^2}{n_\nu + \sigma^2}\right)^2 \lambda_\nu \\ &+ \sigma^2 \sum_{\nu=1}^N \frac{\lambda_\nu n_\nu}{(n_\nu + \sigma^2)^2} + \sum_{\nu=N+1}^\infty \lambda_\nu \\ &= \sigma^2 \sum_{\nu=1}^N \frac{\lambda_\nu}{(n_\nu + \sigma^2)} + \sum_{\nu=N+1}^\infty \lambda_\nu. \end{aligned} \tag{12.2.5}$$

Ignoring the requirement that the n_ν be integers we have that $\sum_{\nu=1}^N \lambda_\nu/(n_\nu + \sigma^2)$ is minimized over $n_1, \ldots, n_N$ subject to $\sum_{\nu=1}^N n_\nu = n$ when $(n_\nu + \sigma^2)$ is proportional to $\lambda_\nu^{1/2}$. This gives

$$n_\nu = \left(\frac{n + N\sigma^2}{\sum_{\mu=1}^N \lambda_\mu^{1/2}}\right) \lambda_\nu^{1/2} - \sigma^2$$

and (12.2.5) becomes

$$E_f E\|f_{\sigma^2/n}\|^2 = \sigma^2 \frac{(\sum_{\mu=1}^N \lambda_\mu^{1/2})^2}{n + N\sigma^2} + \sum_{N+1}^\infty \lambda_\mu^2.$$

The optimal N is then the greatest integer for which

$$\sigma^2 \frac{(\sum_{\mu=1}^N \lambda_\mu^{1/2})\lambda_N^{1/2}}{n + N\sigma^2} \leq \lambda_N$$

If it is only assumed that $f \in \mathcal{H}_R$, or $f \in \mathcal{C}_\rho$, then an optimal design would depend on the strategy for choosing λ, among other things. It appears plausible that such designs will involve replications of eigenfunctions of the r.k., however.

The nature of optimal designs when $\|f\|^2$ is replaced by $\|P_1 f\|^2$ in (12.2.3) is an open question.

Bibliography

M. ABRAMOWITZ AND I. STEGUN (1965), *Handbook of Mathematical Functions with Formulas, Graphs and Mathematical Tables*, U. S. Government Printing Office, Washington, D.C.

R. ADAMS (1975), *Sobolev Spaces*, Academic Press, New York.

G. AGARWAL AND W. STUDDEN (1980), *Asymptotic integrated mean square error using least squares and bias minimizing splines*, Ann. Statist., 8, pp. 1307–1325.

N. AKHIEZER AND I. GLAZMAN (1963), *Theory of Linear Operators in Hilbert Space*, Ungar, New York.

D. ALLEN (1974), *The relationship between variable selection and data augmentation and a method for prediction*, Technometrics, 16, pp. 125–127.

N. ALTMAN (1987), *Smoothing data with correlated errors*, Tech. Report 280, Department of Statistics, Stanford University, Stanford, CA.

T. ANDERSON (1958), *An Introduction to Multivariate Analysis*, John Wiley, New York.

J. ANDERSON AND A. SENTHILSELVAN (1982), *A two-step regression model for hazard functions*, Appl. Statist., 31, pp. 44–51.

R. ANDERSON, F. DEHOOG, AND M. LUKAS (1980), *The Application and Numerical Solution of Integral Equations*, Sijthoff and Noordhoff, Alphen aan den Rijn, The Netherlands.

R. ANDERSSEN AND P. BLOOMFIELD (1974), *A time series approach to numerical differentiation*, Technometrics, 16, pp. 69–75.

R. ANDERSSEN AND A. JAKEMAN (1975), *Abel type integral equations in stereology, ii: computational methods of solution and the random spheres approximation*, J. Microscopy, 105, pp. 135–153.

D. ANDREWS (1988), *Asymptotic optimality of* GC_L*, cross-validation, and* GCV *in regression with heteroscedastic errors*, Cowles Foundation, Yale University, New Haven, CT, manuscript.

C. ANSLEY AND R. KOHN (1987), *Efficient generalized cross-validation for state space models*, Biometrika, 74, pp. 139–148.

C. ANSLEY AND W. WECKER (1981), *Extensions and examples of the signal extraction approach to regression*, in Proc. of ASA-CENSUS-NBER Conference on Applied Time Series, Washington, D.C.

N. ARONSZAJN (1950), *Theory of reproducing kernels*, Trans. Amer. Math. Soc., 68, pp. 337–404.

M. ATHAVALE AND G. WAHBA (1979), *Determination of an optimal mesh for a collocation-projection method for solving two-point boundary value problems*, J. Approx. Theory, 28, pp. 38–48.

G. BACKUS AND F. GILBERT (1970), *Uniqueness in the inversion of inaccurate gross earth data*, Philos. Trans. Roy. Soc. London Ser. A., 266, pp. 123–192.

C. BAKER AND G. MILLER, EDS. (1982), *Treatment of Integral Equations by Numerical Methods*, Academic Press, London.

D. BARRY (1983), *Nonparametric Bayesian Regression*, Ph.D. thesis, Yale University, New Haven, CT.

D. BARRY (1986), *Nonparametric Bayesian regression*, Ann. Statist., 14, pp. 934–953.

D. BARRY AND J. HARTIGAN (1988), *An omnibus test for departures from constant mean*, manuscript.

D. BATES AND G. WAHBA (1982), *Computational methods for generalized cross-validation with large data sets*, in Treatment of Integral Equations by Numerical Methods, C. Baker and G. Miller, eds., Academic Press, London.

D. BATES AND G. WAHBA (1983), *A truncated singular value decomposition and other methods for generalized cross-validation*, Tech. Report 715, Department of Statistics, University of Wisconsin, Madison, WI.

D. BATES AND D. WATTS (1988), *Nonlinear Regression Analysis and Its Applications*, John Wiley, New York.

D. BATES, M. LINDSTROM, G. WAHBA, AND B. YANDELL (1987), GCVPACK-*routines for generalized cross validation*, Comm. Statist. B—Simulation Comput., 16, pp. 263–297.

L. BONEVA, D. KENDALL, AND I. STEFANOV (1971), *Spline transformations*, J. Roy. Stat. Soc. Ser. A, 33, pp. 1–70.

A. BUJA, T. HASTIE, AND R. TIBSHIRANI (1989), *Linear smoothers and additive models*, Ann. Statist., 17, pp. 453–555.

P. BURMAN (1985), *Estimation of generalized additive models*, Ph.D. thesis, Rutgers University, New Brunswick, NJ.

H. CHEN (1988), *Convergence rates for parametric components in a partly linear model*, Ann. Statist., 16, pp. 136–146.

Z. CHEN (1987), *A stepwise approach for the purely periodic interaction spline model*, Comm. Statist. A—Theory Methods, 16, pp. 877–895.

Z. CHEN (1989), *Interaction spline models*, Ph.D. thesis, Department of Statistics, University of Wisconsin, Madison, WI.

Z. CHEN, C. GU, AND G. WAHBA (1989), *Comments to "Linear Smoothers and Additive Models," by Buja, Hastie and Tibshirani*, Ann. Statist., 17, pp. 515–521.

C. CHUI (1988), *Multivariate Splines*, Society for Industrial and Applied Mathematics, Philadelphia, PA.

B. COOK (1963), *Least structure solution of photonuclear yield functions*, Nuclear Instruments and Methods, 24, pp. 256–268.

R. COURANT AND D. HILBERT (1965), *Methods of Mathematical Physics*, vol. 1, Interscience, New York.

D. D. COX (1983), *Asymptotics for m type smoothing splines*, Ann. Statist. , 11, pp. 530–551.

D. D. COX (1984), *Multivariate smoothing spline functions*, SIAM J. Numer. Anal., 21, pp. 789–813.

D. D. COX (1988), *Approximation of method of regularization estimators*, Ann. Statist., 16, pp. 694–713.

D. D. COX AND E. KOH (1986), *A smoothing spline based test of model adequacy in polynomial regression*, Tech. Report 787, Department of Statistics, University of Wisconsin, Madison, WI.

D. COX AND F. O'SULLIVAN (1989a), *Asymptotic analysis of penalized likelihood and related estimators*, Tech. Report 168, Department of Statistics, University of Washington, Seattle, WA.

D. COX AND F. O'SULLIVAN (1989b), *Generalized nonparametric regression via penalized likelihood*, Tech. Report 1970, Department of Statistics, University of Washington, Seattle, WA.

D. D. COX, E. KOH, G. WAHBA, AND B. YANDELL (1988), *Testing the (parametric) null model hypothesis in (semiparametric) partial and generalized spline models*, Ann. Statist., 16, pp. 113–119.

H. CRAMER AND M. LEADBETTER (1967), *Stationary and Related Stochastic Processes*, John Wiley, New York.

P. CRAVEN AND G. WAHBA (1979), *Smoothing noisy data with spline functions: estimating the correct degree of smoothing by the method of generalized cross-validation*, Numer. Math., 31, pp. 377–403.

N. CRESSIE AND R. HORTON (1987), *A robust-resistant spatial analysis of soil water infiltration*, Water Resources Res., 23, pp. 911–917.

J. CRUMP AND J. SEINFELD (1982), *A new algorithm for inversion of aerosol size distribution data*, Aerosol Sci. Tech., 1, pp. 15–34.

A. DAVIES AND R. ANDERSSEN (1985), *Improved estimates of statistical regularization parameters in Fourier differentiation and smoothing*, Tech. Report CMA-R01-85, Center for Mathematical Analysis, Australian National University, Canberra, Australia.

C. DEBOOR (1978), *A Practical Guide to Splines*, Springer-Verlag, New York.

C. DEBOOR AND R. LYNCH (1966), *On splines and their minimum properties*, J. Math. Mech., 15, pp. 953–969.

P. DELFINER (1975), *Linear estimation of non stationary spatial phenomena*, in Advanced Geostatistics in the Mining Industry, M. Guarascio, M. David, and C. Huijbregts, eds., D. Reidel, Dordrecht, Holland, pp. 44–68.

F. DELVOS AND H. POSDORF (1977), *Nth-order blending*, in Constructive Theory of Functions of Several Variables, Lecture Notes in Mathematics 571, W. Schempp and K. Zeller, eds., Springer-Verlag, Berlin, New York, pp. 53–64.

L. DENBY (1986), *Smooth regression function*, Tech. Report 26, AT&T Bell Laboratories, Murray Hill, NJ.

R. DEVEAUX AND J. STEELE (1989), *ACE guided transformation method for estimation of the coefficient of soil water diffusivity*, Technometrics, 31, pp. 91–98.

C. DOLPH AND M. WOODBURY (1952), *On the relations between Green's functions and covariance of certain stochastic processes*, Trans. Amer. Math. Soc., 72, pp. 519–550.

J. DONGARRA, J. BUNCH, C. MOLER, AND G. STEWART (1979), *Linpack Users' Guide*, Society for Industrial and Applied Mathematics, Philadelphia, PA.

D. DONOHO, AND I. JOHNSTONE (1989), *Projection-based approximation and a duality with kernel methods*, Ann. Statist., 17, pp. 58–106.

J. DUCHON (1975), *Fonctions splines et vecteurs aleatoires*, Tech. Report 213, Seminaire d'Analyse Numerique, Universite Scientifique et Medicale, Grenoble.

J. DUCHON (1976), *Fonctions-spline et esperances conditionnelles de champs gaussiens*, Ann. Sci. Univ. Clermont Ferrand II Math., 14, pp. 19–27.

J. DUCHON (1977), *Splines minimizing rotation-invariant semi-norms in Sobolev spaces*, in Constructive Theory of Functions of Several Variables, Springer-Verlag, Berlin, pp. 85–100.

J. DUCHON (1978), *Sur l'erreur d'interpolation des fonctions du plusieurs variables par les D sup m-splines*, R.A.I.R.O Anal. Numer., 12, pp. 325–334.

N. DYN, AND G. WAHBA (1982), *On the estimation of functions of several variables from aggregated data*, SIAM J. Math. Anal., 13, pp. 134–152.

N. DYN, G. WAHBA, AND W. WONG (1979), *Comment on "Smooth pychnophylactic interpolation for geographical regions* by W. Tobler," J. Amer. Statist. Assoc., 74, pp. 530–535.

B. EFRON (1982), *The Jackknife, the Bootstrap and Other Resampling Plans*, Society for Industrial and Applied Mathematics, Philadelphia, PA.

B. EFRON AND R. TIBSHIRANI (1986), *Bootstrap methods for standard errors, confidence intervals, and other measures of statistical accuracy*, Statist. Sci., 1, pp. 54–77.

L. ELDEN (1977), *Algorithms for the regularization of ill-conditioned least squares problems*, BIT, 17, pp. 134–145.

L. ELDEN (1984), *A note on the computation of the generalized cross-validation function for ill-conditioned least squares problems*, BIT, 24, pp. 467–472.

T. ELFVING AND L. ANDERSSON (1986), *An algorithm for computing constrained smoothing spline functions*, Tech. Report MAT-R-86-07, Linkoping University Institute of Technology, Linkoping, Sweden.

R. ENGLE, C. GRANGER, J. RICE, AND A. WEISS (1986), *Semiparametric estimates of the relation between weather and electricity sales*, J. Amer. Statist. Soc., 81, pp. 310–320.

R. EUBANK (1984), *The hat matrix for smoothing splines*, Statist. & Probab. Lett., 2, pp. 9–14.

R. EUBANK (1985), *Diagnostics for smoothing splines*, J. Roy. Statist. Soc. Ser. B, 47, pp. 332–341.

R. EUBANK (1986), *A note on smoothness priors and nonlinear regression*, J. Amer. Statist. Assoc., 81, pp. 514–517.

R. FRANKE (1979), *A critical comparison of some methods for interpolation of scattered data*, Tech. Report NPS-53-79-003, Naval Postgraduate School, Monterey, CA.

W. FREEDEN (1981), *On spherical spline interpolation and approximation*, Math. Meth. Appl. Sci., 3, pp. 551–575.

J. FRIEDMAN (1989), *Multivariate adaptive regression splines*, Ann. Statist., to appear.

J. FRIEDMAN AND B. SILVERMAN (1989), *Flexible parsimonious smoothing and additive modeling*, Technometrics, 3, pp. 3–21.

J. FRIEDMAN, E. GROSSE, AND W. STUETZLE (1983), *Multidimensional additive spline approximation*, SIAM J. Sci. Statist. Comput., 4, pp. 291–301.

S. FRITZ, D. WARK, J. FLEMING, W. SMITH, H. JACOBOWITZ, D. HILLEARY, AND J. ALISHOUSE (1972), *Temperature sounding from satellites*, Tech. Report NESS 59, National Oceanic and Atmospheric Administration, Washington, D.C.

H. GAMBER (1979), *Choice of an optimal shape parameter when smoothing noisy data*, Comm. Statist. A—Theory Methods, 8, pp. 1425–1436.

P. GILL, W. MURRAY, AND M. WRIGHT (1981), *Practical Optimization*, Academic Press, New York.

P. GILL, N. GOULD, W. MURRAY, M. SAUNDERS, AND M. WRIGHT (1982), *A range-space method for quadratic programming problems with bounds and general constraints*, Tech. Report SOL 82-15, Department of Operations Research, Stanford University, Stanford, CA.

D. GIRARD (1987a), *A fast 'Monte-Carlo cross-validation' procedure for large least squares problems with noisy data*, Tech. Report RR 687-M, Informatique et Mathematiques Appliquees de Grenoble, Grenoble.

D. GIRARD (1987b), *Un algorithme rapide pour le calcul de la trace de l'universe d'une grande matrice*, Tech. Report RR 665-M, Informatique et Mathematiques Appliquees de Grenoble, Grenoble.

D. GIRARD (1987c), *Optimal regularized reconstruction in computerized tomography*, SIAM J. Sci. Statist. Comput., 8, pp. 934–950.

M. GOLOMB AND H. WEINBERGER (1959), *Optimal approximation and error bounds*, in Proc. Symp. on Numerical Approximation, R. Langer, ed., University of Wisconsin Press, Madison, WI, pp. 117–190.

G. GOLUB AND C. V. LOAN (1983), *Matrix Computations*, Johns Hopkins University Press, Baltimore, MD.

G. GOLUB, M. HEATH, AND G. WAHBA (1979), *Generalized cross validation as a method for choosing a good ridge parameter*, Technometrics, 21, pp. 215–224.

P. GREEN (1985), *Linear models for field trials, smoothing and cross validation*, Biometrika, 72, pp. 527–537.

P. GREEN (1987), *Penalized likelihood for general semi-parametric regression models*, Internat. Statist. Rev., 55, pp. 245–259.

P. GREEN AND B. YANDELL (1985), *Semi-parametric generalized linear models*, Lecture Notes in Statistics 32, Springer-Verlag, Berlin, New York, pp. 44–55.

P. GREEN, C. JENNISON, AND A. SEHEULT (1983), *Comments to "Nearest neighbor (NN) analysis of field experiments" by Wilkinson, et al.*, J. Roy. Statist. Soc. Ser. B, 45, pp. 193–195.

P. GREEN, C. JENNISON, AND A. SEHEULT (1985), *Analysis of field experiments by least squares smoothing*, J. Roy. Statist. Soc. Ser. B, 47, pp. 299–315.

T GREVILLE, ED. (1968), *Theory and Application of Spline Functions*, University of Wisconsin Press, Madison, WI.

C. GROETSCH (1984), *The Theory of Tikhonov Regularization for Fredholm Equations of the*

First Kind, Pitman, Boston, MA.

C. GU (1988), *personal communication.*

C. GU (1989a), *Generalized spline models: A convenient algorithm for optimal smoothing*, Tech. Report 853, Department of Statistics, University of Wisconsin, Madison, WI.

C. GU (1989b), RKPACK *and its applications: fitting smoothing spline models*, Tech. Report 857, Department of Statistics, University of Wisconsin, Madison, WI.

C. GU AND G. WAHBA (1988), *Minimizing* GCV/GML *scores with multiple smoothing parameters via the Newton method*, Tech. Report 847, Department of Statistics, University of Wisconsin, Madison, WI; SIAM J. Sci. Statist., to appear.

C. GU, D. BATES, Z. CHEN, AND G. WAHBA (1989), *The computation of* GCV *functions through Householder tridiagonalization with application to the fitting of interaction spline models*, SIAM J. Matrix Anal., 10, pp. 459–480.

J. HAJEK (1962a), *On linear statistical problems in stochastic processes*, Czech. Math. J., 87, pp. 404–444.

J. HAJEK (1962b), *On a property of normal distributions of any stochastic process*, Selected Trans. Math. Statist. Probab., 1, pp. 245–252.

P. HALL AND D. TITTERINGTON (1987), *Common structure of techniques for choosing smoothing parameters in regression problems*, J. Roy. Statist. Soc. Ser. B, 49, pp. 184–198.

P. HALMOS (1957), *Introduction to Hilbert Space and the Theory of Spectral Multiplicity*, Chelsea, New York.

W. HARDLE, P. HALL, AND S. MARRON (1988), *How far are automatically chosen smoothing parameters from their optimum?*, J. Amer. Statist. Assoc., 83, pp. 86–95; Rejoinder, pp. 100–101.

T. HASTIE AND R. TIBSHIRANI (1986), *Generalized additive models*, Statist. Sci., 1, pp. 297–318.

N. HECKMAN (1986), *Spline smoothing in a partly linear model*, J. Roy. Statist. Soc. Ser. B, 48, pp. 244–248.

T. HIDA (1960), *Canonical representations of Gaussian processes and their applications*, Mem. Colloq. Sci. Kyoto Univ. Ser A. Math, 32, pp. 109–155.

R. HOFFMAN (1984), *SASS wind ambiguity removal by direct minimization II. Effect of smoothing and dynamical constraints*, Monthly Weather Review, 112, pp. 1829–1852.

R. HOFFMAN (1985), *Using smoothness constraints in retrievals*, in Advances in Remote Sensing Retrieval Methods, A. Deepak, H. Fleming, and M. Chahine, eds., pp. 411–436, A. Deepak Publishing, Hampton, VA.

K. HOLLIG (1986), *Multivariate splines*, in Approximation Theory, C. de Boor, ed., American Mathematical Society, Providence, RI.

A. HOLLINGSWORTH (1989), *personal communication.*

P. HUBER (1985), *Projection pursuit*, Ann. Statist., 13, pp. 435–525.

H. HUDSON (1974), *Empirical Bayes Estimation*, Tech. Report 58, Department of Statistics, Stanford University, Stanford, CA.

C. HURVICH (1985), *Data driven choice of a spectrum estimate: extending the applicability of cross-validation methods*, J. Amer. Statist. Assoc., 80, pp. 933–940.

M. HUTCHINSON (1984), *A summary of some surface fitting and contouring programs for noisy data*, Tech. Report ACT 84/6, CSIRO Division of Mathematics and Statistics, Canberra, Australia.

M. HUTCHINSON (1985), *Algorithm 642, a fast procedure for calculating minimum cross validation cubic smoothing splines*, ACM Trans. Math. Software, 12, pp. 150–153.

M. HUTCHINSON (1989), *A stochastic estimator for the trace of the influence matrix for Laplacian smoothing splines*, Commun. Statist. Simul. Comp., 18, pp. 1059–1076.

M. HUTCHINSON AND R. BISCHOF (1983), *A new method for estimating the spatial distribution of mean seasonal and annual rainfall applied to the Hunter Valley, New South Wales*, Aust. Met. Mag., 31, pp. 179–184.

M. HUTCHINSON AND F. DEHOOG (1985), *Smoothing noisy data with spline functions*, Numer. Math., 47, pp. 99–106.

IMSL (1986), *International Mathematical and Statistical Library*, Houston, TX.

A. IVIC (1985), *The Riemann Zeta-Function*, John Wiley, New York.

I. JOHNSTONE AND B. SILVERMAN (1988), *Speed of estimation in positron emission tomography*, Tech. Report 290, Department of Statistics, Stanford University, Stanford, CA.

A. JOURNEL AND C. HUIJBREGTS (1978), *Mining Geostatistics*, Academic Press, Orlando, FL.

S. KARLIN (1968), *Total Positivity*, Stanford University Press, Stanford, CA.

G. KIMELDORF AND G. WAHBA (1970a), *A correspondence between Bayesian estimation of stochastic processes and smoothing by splines*, Ann. Math. Statist., 41, pp. 495–502.

G. KIMELDORF AND G. WAHBA (1970b), *Spline functions and stochastic processes*, Sankhya Ser. A, 32, Part 2, pp. 173–180.

G. KIMELDORF AND G. WAHBA (1971), *Some results on Tchebycheffian spline functions*, J. Math. Anal. Appl., 33, pp. 82–95.

R. KOHN AND C. ANSLEY (1987), *A new algorithm for spline smoothing based on smoothing a stochastic process*, SIAM J. Sci. Statist. Comput., 8, pp. 33–48.

C. KRAVARIS AND J. SEINFELD (1985), *Identification of parameters in distributed parameter systems by regularization*, SIAM J. Control Optim., 23, pp. 217–241.

P. LAURENT AND F. UTRERAS (1986), *Optimal smoothing of noisy broken data*, Approx. Theory Appl., 2, pp. 71–94.

D. LEGLER, I. NAVOW, AND J. O'BRIEN (1989), *Objective analysis of pseudostress over the Indian Ocean using a direct-minimization approach*, Monthly Weather Review, 17, pp. 709–720.

T. LEONARD (1982), *An empirical Bayesian approach to the smooth estimation of unknown functions*, Tech. Report 2339, Mathematics Research Center, University of Wisconsin, Madison, WI.

K. C. LI (1985), *From Stein's unbiased risk estimates to the method of generalized cross-validation*, Ann. Statist., 13, pp. 1352–1377.

K. C. LI (1986), *Asymptotic optimality of C_L and generalized cross validation in ridge regression with application to spline smoothing*, Ann. Statist., 14, pp. 1101–1112.

K. C. LI (1987), *Asymptotic optimality for C_p , C_L , cross-validation and generalized cross validation: discrete index set*, Ann. Statist., 15, pp. 958–975.

P. LONNBERG AND A. HOLLINGSWORTH (1986), *The statistical structure of short-range forecast errors as determined from radiosonde data, Part II: The covariance of height and wind errors*, Tellus, 38A, pp. 137–161.

M. LUKAS (1981), *Regularization of linear operator equations*, Ph.D. thesis, Australian National University, Canberra, ACT.

T. LYCHE AND L. SCHUMAKER (1973), *Computation of smoothing and interpolating natural splines via local bases*, SIAM J.Numer. Anal., 10, pp. 1027–1038.

C. MALLOWS (1973), *Some comments on C_p*, Technometrics, 15, pp. 661–675.

L. MANSFIELD (1972), *On the variational characterization and convergence of bivariate splines*, Numer. Math., 20, pp. 99–114.

G. MATHERON (1973), *The intrinsic random functions and their applications*, Adv. Appl. Probab., 5, pp. 439–468.

P. MCCULLAGH AND J. NELDER (1983), *Generalized Linear Models*, Chapman and Hall, London.

J. MEINGUET (1979), *Multivariate interpolation at arbitrary points made simple*, J. Appl. Math. Phys. (ZAMP), 30, pp. 292–304.

A. MELKMAN AND C. MICCHELLI (1978), *Spline spaces are optimal for L_2 n-widths*, Illinois J. Math., 22, pp. 541–564.

J. MENDELSSON AND J. RICE (1982), *Deconvolution of microfluorometric histograms with B-splines*, J. Amer. Statist. Soc., 77, pp. 748–753.

P. MERZ (1980), *Determination of adsorption energy distribution by regularization and a characterization of certain adsorption isotherms*, J. Comput. Phys., 38, pp. 64–85.

C. MICCHELLI AND G. WAHBA (1981), *Design problems for optimal surface interpolation*, in Approximation Theory and Applications, Z. Ziegler, ed., Academic Press, New York, pp. 329–348.

F. MOSTELLER AND D. WALLACE (1963), *Inference in an authorship problem. A comparative study of discrimination methods applied to the authorship of the disputed Federalist papers*, J. Amer. Statist. Assoc., 58, pp. 275–309.

I. NAIMARK (1967), *Linear Differential Operators*, Vol. 2, Ungar, New York.

M. NASHED AND G. WAHBA (1974), *Generalized inverses in reproducing kernel spaces: An approach to regularization of linear operator equations*, SIAM J. Math. Anal., 5, pp. 974–987.

J. NELDER AND R. WEDDERBURN (1972), *Generalized linear models*, J. Roy. Statist. Soc. Ser. A, 135, pp. 370–384.

D. NYCHKA (1983), *personal communication.*

D. NYCHKA (1986a), *The average posterior variance of a smoothing spline and a consistent estimate of the mean square error*, Tech. Report 168, The Institute of Statistics, North Carolina State University, Raleigh, NC.

D. NYCHKA (1986b), *A frequency interpretation of Bayesian "confidence" intervals for smoothing splines*, Tech. Report 169, The Institute of Statistics, North Carolina State University, Raleigh, NC; J. Amer. Statist. Assoc., to appear.

D. NYCHKA (1988), *Confidence intervals for smoothing splines*, J. Amer. Statist. Assoc., 83, pp. 1134–1143.

D. NYCHKA AND D. COX (1989), *Convergence rates for regularized solutions of integral equations from discrete noisy data*, Ann. Statist., 17, pp. 556–572.

D. NYCHKA, G. WAHBA, S. GOLDFARB, AND T. PUGH (1984), *Cross-validated spline methods for the estimation of three dimensional tumor size distributions from observations on two dimensional cross sections*, J. Amer. Stat. Assoc., 79, pp. 832–846.

A. O'HAGAN (1976), *On posterior joint and marginal modes*, Biometrika, 63, pp. 329–333.

F. O'SULLIVAN (1983), *The analysis of some penalized likelihood estimation schemes*, Ph.D. thesis, Tech. Report 726, Department of Statistics, University of Wisconsin, Madison, WI.

F. O'SULLIVAN (1985a), *Inverse problems: bias, variability and risk assessment*, Department of Statistics, University of California, Berkeley, CA.

F. O'SULLIVAN (1985b), *Comments on "Some aspects of the spline smoothing approach to non-parametric regression curve fitting," by B. Silverman*, J. Roy. Statist. Soc. B, 47, pp. 39–40.

F. O'SULLIVAN (1986a), *A statistical perspective on ill-posed inverse problems*, Statist. Sci., 1, pp. 502–527.

F. O'SULLIVAN (1986b), *Estimation of densities and hazards by the method of penalized likelihood*, Tech. Report 58, Department of Statistics, University of California, Berkeley, CA.

F. O'SULLIVAN (1986c), *Evaluating the performance of an inversion algorithm*, in Contemporary Mathematics 59, S. Marron, ed., American Mathematical Society, Providence, RI, pp. 53–61.

F. O'SULLIVAN (1987a), *Fast computation of fully automated log- density and log-hazard estimators*, SIAM J. Sci. Statist. Comput., 9, pp. 363–379.

F. O'SULLIVAN (1987b), *Constrained non-linear regularization with application to some system identification problems*, Tech. Report 99, Department of Statistics, University of California, Berkeley, Berkeley, CA.

F. O'SULLIVAN (1988a), *Parameter estimation in parabolic and hyperbolic equations*, Tech. Report 127, Department of Statistics, University of Washington, Seattle, WA.

F. O'SULLIVAN (1988b), *Nonparametric estimation of relative risk using splines and cross-validation*, SIAM J. Sci. Statist. Comput., 9, pp. 531–542.

F. O'SULLIVAN AND G. WAHBA (1985), *A cross validated Bayesian retrieval algorithm for non-linear remote sensing*, J. Comput. Phys., 59, pp. 441–455.

F. O'SULLIVAN, B. YANDELL, AND W. RAYNOR (1986), *Automatic smoothing of regression functions in generalized linear models*, J. Amer. Statist. Assoc., 81, pp. 96–103.

F. O'SULLIVAN AND T. WONG (1987), *Determining a function diffusion coefficient in the heat equation*, Tech. Report 98, Department of Statistics, University of California, Berkeley, Berkeley, CA.

E. PARZEN (1962), *An approach to time series analysis*, Ann. Math. Statist., 32, pp. 951–989.

E. PARZEN (1963), *Probability density functionals and reproducing kernel Hilbert spaces*, in Proc. Symposium on Time Series Analysis, M. Rosenblatt, ed., John Wiley, New York, pp. 155–169.

E. PARZEN (1970), *Statistical inference on time series by rkhs methods*, in Proc. 12th Biennial Seminar, R. Pyke, ed., Canadian Mathematical Congress, Montreal, Canada, pp. 1-37.

D. PHILLIPS (1962), *A technique for the numerical solution of certain integral equations of the first kind*, J. Assoc. Comput. Mach., 9, pp. 84–97.

L. PLASKOTA (1989), *On average case complexity of linear problems with noisy information*, Tech. Report, Institute of Informatics, University of Warsaw, Warsaw, Poland.

P. PRENTER (1975), *Splines and Variational Methods*, John Wiley, New York.

D. RAGOZIN (1983), *Error bounds for derivative estimates based on spline smoothing of exact or noisy data*, J. Approx. Theory, 37, pp. 335–355.

C. REINSCH (1971), *Smoothing by spline functions II*, Numer. Math., 16, pp. 451–454.

J. RICE (1986), *Convergence rates for partially splined models*, Statist. Probab. Lett., 4, pp. 203–208.

J. RICE AND M. ROSENBLATT (1983), *Smoothing splines: regression, derivatives and deconvolution*, Ann. Statist., 11, pp. 141–156.

F. RIESZ AND B. SZ.-NAGY (1955), *Functional Analysis*, Ungar, New York.

M. ROSENBLATT, ED. (1963), *Proceedings of the Symposium on Time Series Analysis, held at Brown University, 1962*, John Wiley, New York.

J. SACKS AND D. YLVISAKER (1969), *Designs for regression problems with correlated errors iii*, Ann. Math. Statist., 41, pp. 2057–2074.

G. SANSONE (1959), *Orthogonal Functions*, Interscience, New York.

I. SCHOENBERG (1942), *Positive definite functions on spheres*, Duke Math. J., 9, pp. 96–108.

I. SCHOENBERG (1964a), *Spline functions and the problem of graduation*, Proc. Nat. Acad. Sci. U.S.A., 52, pp. 947–950.

I. SCHOENBERG (1964b), *On interpolation by spline functions and its minimum properties*, Internat. Ser. Numer. Anal., 5, pp. 109–129.

I. SCHOENBERG (1968), *Monosplines and quadrature formulae*, in Theory and Application of Spline Functions, T. Greville, ed., University of Wisconsin Press, Madison, WI.

I. SCHOENBERG, ED. (1969), *Proceedings of the Symposium on Approximation, with Special Emphasis on Splines*, Academic Press, New York.

M. SCHULTZ (1973a), *Spline Analysis*, Prentice-Hall, Englewood Cliffs, NJ.

M. SCHULTZ (1973b), *Error bounds for a bivariate interpolation scheme*, J. Approx. Theory, 8, pp. 184–193.

L. SCHUMAKER (1981), *Spline Functions*, John Wiley, New York.

L. SCHUMAKER AND F. UTRERAS (1988), *Asymptotic properties of complete smoothing splines and applications*, SIAM J. Sci. Statist. Comput., 9, pp. 24–38.

D. SCOTT (1988), *Comment on "How far are automatically chosen smoothing parameters from their optimum?"*, by Hardle, Hall and Marron, J. Amer. Statist. Assoc., 83, pp. 96–98.

D. SCOTT AND G. TERRELL (1987), *Biased and unbiased cross-validation in density estimation*, J. Amer. Statist. Assoc., 82, pp. 1131–1146.

R. SEAMAN AND M. HUTCHINSON (1985), *Comparative real data tests of some objective analysis methods by withholding*, Aust. Met. Mag., 33, pp. 37–46.

T. SEVERINI AND W. H. WONG (1987), *Convergence rates of maximum likelihood and related estimates in general parameter spaces*, Tech. Report 207, Department of Statistics, The University of Chicago, Chicago, IL.

B. SHAHRARY AND D. ANDERSON (1989), *Optimal estimation of contour properties by cross-validated regularization*, IEEE Trans. Pattern Anal. Mach. Intell., 11, pp. 600–610.

L. SHEPP (1966), *Radon-Nikodym derivatives of Gaussian measures*, Ann. Math. Statist., 37, pp. 321–354.

J. SHIAU (1985), *Smoothing spline estimation of functions with discontinuities*, Ph.D. thesis, Department of Statistics, University of Wisconsin, Madison, also Tech. Report 768.

J. SHIAU (1988), *Efficient algorithms for smoothing spline estimation of functions with and without discontinuities*, in Computer Science and Statistics, 19th Conference on the Interface, E. Wegman, ed., American Statistical Association, Washington, DC.

J. SHIAU AND G. WAHBA (1988), *Rates of convergence of some estimators for a semiparametric model*, Comm. Statist. Simulation and Computation, 17, pp. 1117–1133.

J. SHIAU, G. WAHBA, AND D. JOHNSON (1986), *Partial spline models for the inclusion of tropopause and frontal boundary information*, J. Atmospheric Ocean Tech., 3, pp. 714–725.

R. SHILLER (1984), *Smoothness priors and nonlinear regression*, J. Amer. Statist. Assoc., 79, pp. 609–615.

L. SHURE, R. PARKER, AND G. BACKUS (1982), *Harmonic splines for geomagnetic modelling*, J. Phys. Earth Planetary Interiors, 28, pp. 215–229.

B. SILVERMAN (1978), *Density ratios, empirical likelihood and cot death*, J. Roy. Statist. Soc. Ser. B, 27, pp. 26–33.

B. SILVERMAN (1982), *On the estimation of a probability density function by the maximum penalized likelihood method*, Ann. Statist., 10, pp. 795–810.

B. SILVERMAN (1984), *A fast and efficient cross validation method for smoothing parameter choice in spline regression*, J. Amer. Statist. Assoc., 79, pp. 584–589.

B. SILVERMAN (1985), *Some aspects of the spline smoothing approach to non-parametric regression curve fitting*, J. Roy. Statist. Soc. Ser. B, 46, pp. 1–52.

B. SMITH, J. BOYLE, J. DONGARRA, B. GARBOW, Y. IKEBE, V. KLEMA, AND C. MOLER (1976), *Matrix Eigensystem Routines*-EISPACK *Guide*, Springer-Verlag, New York.

P. SPECKMAN (1985), *Spline smoothing and optimal rates of convergence in nonparametric regression*, Ann. Statist., 13, pp. 970–983.

P. SPECKMAN (1988), *Kernel smoothing in partial linear models*, J. Roy. Statist. Soc. Ser. B, 50, pp. 413–436.

J. STANFORD (1979), *Latitudinal-wavenumber power spectra of stratospheric temperature fluctuations*, J. Atmospheric Sci., 36, pp. 921–931.

M. STEIN (1985), *Asymptotically efficient spatial interpolation with a misspecified covariance function*, Tech. Report 186, Department of Statistics, University of Chicago, Chicago, IL.

M. STEIN (1987a), *Minimum norm quadratic estimation of spatial variograms*, J. Amer. Statist. Assoc., 82, pp. 765–772.

M. STEIN (1987b) *Uniform asymptotic optimality of linear predictions of a random field using an incorrect second-order structure*, Tech. Report, Department of Statistics, University of Chicago, Chicago, IL.

D. STEINBERG (1983), *Bayesian models for response surfaces of uncertain functional form*, Tech. Report 2474, Mathematics Research Center, University of Wisconsin, Madison, WI.

D. STEINBERG (1984a), *Bayesian models for response surfaces* I: *The equivalence of several models*, Tech. Report 2682, Mathematics Research Center, University of Wisconsin, Madison, WI.

D. STEINBERG (1984b), *Bayesian models for response surfaces II: Estimating the response surface*, Tech. Report 2683, Mathematics Research Center, University of Wisconsin, Madison, WI.

C. STONE (1985), *Additive regression and other nonparametric models*, Ann. Statist., 13, pp. 689–705.

C. STONE (1986), *The dimensionality reduction principle for generalized additive models*, Ann. Statist., 14.

P. SWARZTRAUBER (1981), *The approximation of vector-functions and their derivatives on the sphere*, SIAM J. Numer. Anal., 18, pp. 191–210.

R. TAPIA AND J. THOMPSON (1978), *Nonparametric Probability Density Estimation*, Johns Hopkins University Press, Baltimore, MD.

C. THOMAS-AGNAN (1987), *Statistical curve fitting by Fourier techniques*, Ph.D. thesis, University of California, Los Angeles, CA.

A. TIKHONOV (1963), *Solution of incorrectly formulated problems and the regularization method*, Soviet Math. Dokl., 4, pp. 1035–1038.

A. TIKHONOV AND V. GONCHARSKY, EDS. (1987), *Ill-Posed Problems in the Natural Sciences*, Mir, Moscow.

D. TITTERINGTON (1985), *Common structure of smoothing techniques in statistics*, Internat. Statist. Rev., 53, pp. 141–170.

F. UTRERAS (1978), *Quelques resultats d'optimalite pour la methode de validation croissee*, Tech. Report 301, Seminaire d'Analyse Numerique.

F. UTRERAS (1979), *Cross-validation techniques for smoothing spline functions in one or two dimensions*, in Smoothing Techniques for Curve Estimation, T. Gasser and M. Rosenblatt, eds., Springer-Verlag, Heidelberg, pp. 196–231.

F. UTRERAS (1980), *Sur le choix du parametre d'adjustement dans le lissage par fonctions spline*, Numer. Math., 34, pp. 15–28.

F. UTRERAS (1981a), *On computing robust splines and applications*, SIAM J. Sci. Statist. Comput., 2, pp. 153–163.

F. UTRERAS (1981b), *Optimal smoothing of noisy data using spline functions*, SIAM J. Sci. Statist. Comput., 2, pp. 349–362.

F. UTRERAS (1983), Natural spline functions: their associated eigenvalue problem, Numer. Math., 42, pp. 107–117.

F. UTRERAS (1985), *Smoothing noisy data under monotonicity constraints, existence, characterization and convergence rates*, Numer. Math., 47, pp. 611–625.

F. UTRERAS (1988a), *Convergence rates for multivariate smoothing spline functions*, J. Approx. Theory, 52, pp. 1–27.

F. UTRERAS (1988b), *On generalized cross-validation for multivariate smoothing spline functions*, SIAM J. Sci. Statist. Comput., 8, pp. 630–643.

M. VILLALOBOS AND G. WAHBA (1987), *Inequality constrained multivariate smoothing splines with application to the estimation of posterior probabilities*, J. Amer. Statist. Assoc., 82, pp. 239–248.

C. VOGEL (1986), *Optimal choice of a truncation level for the truncated* SVD *solution of linear first kind integral equations when data are noisy*, SIAM J. Numer. Anal., 23, pp. 109–117.

M. VON GOLITSCHEK AND L. SCHUMAKER (1987), *Data fitting by penalized least squares*, Algorithms for Approximation II, M. Cox, J. Mason, and A. Watson, eds., to appear.

G. WAHBA (1968), *On the distribution of some statistics useful in the analysis of jointly stationary time series*, Ann. Math. Statist., 39, pp. 1849–1862.

G. WAHBA (1969), *On the numerical solution of Fredholm integral equations of the first kind*, Tech. Report 990, Mathematics Research Center, University of Wisconsin, Madison, WI.

G. WAHBA (1971), *On the regression design problem of Sacks and Ylvisaker*, Ann. Math. Statist., 42, pp. 1035–1043.

G. WAHBA (1973a), *On the minimization of a quadratic functional subject to a continuous family of linear inequality constraints*, SIAM J. Control, 11, pp. 64–79.

G. WAHBA (1973b), *Convergence rates of certain approximate solutions to Fredholm integral equations of the first kind*, J. Approx. Theory, 7, pp. 167–185.

G. WAHBA (1975), *Smoothing noisy data by spline functions*, Numer. Math., 24, pp. 383–393.

G. WAHBA (1976), *On the optimal choice of nodes in the collocation-projection method for solving linear operator equations*, J. Approx. Theory 16, pp. 175–186.

G. WAHBA (1977a), *Practical approximate solutions to linear operator equations when the data are noisy*, SIAM J. Numer. Anal., 14, pp. 651–667.

G. WAHBA (1977b), *Optimal smoothing of density estimates*, in Classification and Clustering, J. VanRyzin, ed., Academic Press, New York, pp. 423–458.

G. WAHBA (1978a), *Comments on "Curve fitting and optimal design for prediction" by A. O'Hagan*, J. Roy. Statist. Soc. Ser. B, 40, pp. 35–36.

G. WAHBA (1978b), *Improper priors, spline smoothing and the problem of guarding against model errors in regression*, J. Roy. Statist. Soc. Ser. B, 40, pp. 364–372.

G. WAHBA (1978c), *Interpolating surfaces: High order convergence rates and their associated designs, with applications to X-ray image reconstruction*, Tech. Report 523, Department of Statistics, University of Wisconsin, Madison, WI.

G. WAHBA (1979a), *Convergence rates of "thin plate" smoothing splines when the data are noisy*, in Smoothing Techniques for Curve Estimation, Lecture Notes in Mathematics, 757, T. Gasser and M. Rosenblatt, eds., Springer-Verlag, Berlin, New York, pp. 232–246.

G. WAHBA (1979b), *Smoothing and ill posed problems*, in Solution Methods for Integral Equations with Applications, M. Golberg, ed., Plenum Press, New York, pp. 183–194.

G. WAHBA (1979c), *How to smooth curves and surfaces with splines and cross-validation*, in Proc. 24th Conference on the Design of Experiments, U.S. Army Research Office, No. 79-2, Research Triangle Park, NC, pp. 167–192.

G. WAHBA (1980a), *Ill posed problems: Numerical and statistical methods for mildly, moderately and severely ill posed problems with noisy data*, Tech. Report 595, Department of Statistics, University of Wisconsin, Madison, WI, prepared for the Proc. International Conference on Ill Posed Problems, M.Z. Nashed, ed.

G. WAHBA (1980b), *Spline bases, regularization, and generalized cross validation for solving approximation problems with large quantities of noisy data*, in Approximation Theory III, W. Cheney, ed., Academic Press, New York, pp. 905–912.

G. WAHBA (1980c), *Automatic smoothing of the log periodogram*, J. Amer. Statist. Assoc., 75, pp. 122–132.

G. WAHBA (1981a) *Cross validation and constrained regularization methods for mildly ill posed problems*, Tech. Report 629, Department of Statistics, University of Wisconsin, Madison, WI.

G. WAHBA (1981b), *Data-based optimal smoothing of orthogonal series density estimates*, Ann. Statist., 9, pp. 146–156.

G. WAHBA (1981c), *A new approach to the numerical inversion of the radon transform with discrete, noisy data*, in Mathematical Aspects of Computerized Tomography, G. Herman and F. Natterer, eds., Springer-Verlag, Berlin, New York.

G. WAHBA (1981d), *Spline interpolation and smoothing on the sphere*, SIAM J. Sci. Statist. Comput., 2, pp. 5–16.

G. WAHBA (1982a), *Erratum: spline interpolation and smoothing on the sphere*, SIAM J. Sci. Statist. Comput., 3, pp. 385–386.

G. WAHBA (1982b), *Vector splines on the sphere, with application to the estimation of vorticity and divergence from discrete, noisy data*, in Multivariate Approximation Theory, Vol.2, W. Schempp and K. Zeller, eds., Birkhauser Verlag, Basel, Boston, Stuttgart, pp. 407–429.

G. WAHBA (1982c), *Constrained regularization for ill posed linear operator equations, with applications in meteorology and medicine*, in Statistical Decision Theory and Related Topics, III, Vol.2, S. Gupta and J. Berger, eds., Academic Press, New York, pp. 383–418.

G. WAHBA (1982), *Variational methods in simultaneous optimum interpolation and initialization*, in the Interaction Between Objective Analysis and Initialization, D. Williamson, ed., Atmospheric Analysis and Prediction Division, National Center for Atmospheric Research, Boulder, CO, pp. 178–185.

G. WAHBA (1983), *Bayesian "confidence intervals" for the cross-validated smoothing spline*, J. Roy. Statist. Soc. Ser. B, 45, pp. 133–150.

G. WAHBA (1984a), *Surface fitting with scattered, noisy data on Euclidean d-spaces and on the sphere*, Rocky Mountain J. Math., 14, pp. 281–299.

G. WAHBA (1984b), *Cross validated spline methods for the estimation of multivariate functions from data on functions, Statistics: An Appraisal*, in Proc. 50th Anniversary Conference Iowa State Statistical Laboratory, Iowa State University Press, Ames, Iowa, pp. 205–235.

G. WAHBA (1984c), *Partial spline models for the semiparametric estimation of functions of several variables*, in Statistical Analysis of Time Series, Proceedings of the Japan U.S. Joint Seminar, Tokyo, pp. 319–329.

G. WAHBA (1984d), *Cross validated spline methods for direct and indirect sensing experiments*, in Statistical Signal Processing, E. Wegman and J. Smith, eds., Marcel Dekker, New York, pp. 179–197.

G. WAHBA (1985a), *Variational methods for multidimensional inverse problems*, in Remote Sensing Retrieval Methods, H. Fleming and M. Chahine, eds., A. Deepak Publishing, Hampton, VA, pp. 385–408.

G. WAHBA (1985b), *Multivariate thin plate spline smoothing with positivity and other linear inequality constraints*, in Statistical Image Processing and Graphics, E. Wegman and D. dePriest, eds., Marcel Dekker, New York, pp. 275–290.

G. WAHBA (1985c), *Comments to P. Huber, Projection pursuit*, Ann. Statist., 13, pp. 518–521.

G. WAHBA (1985d), *Partial spline modelling of the tropopause and other discontinuities*, in Function Estimates, S. Marron, ed., American Mathematical Society, Providence, RI, pp. 125–135.

G. WAHBA (1985e), *A comparison of* GCV *and* GML *for choosing the smoothing parameter in the generalized spline smoothing problem*, Ann. Statist., 13, pp. 1378–1402.

G. WAHBA (1986), *Partial and interaction splines for the semiparametric estimation of functions of several variables*, in Computer Science and Statistics: Proceedings of the 18th Symposium, T. Boardman, ed., American Statistical Association, Washington, DC, pp. 75–80.

G. WAHBA (1987a), *Three topics in ill posed problems*, in Proc. Alpine-U.S. Seminar on Inverse and Ill Posed Problems, H. Engl and C. Groetsch, eds., Academic Press, New York, pp. 37–51.

G. WAHBA (1987b), *Comments to "A statistical perspective on ill posed inverse problems" by F. O'Sullivan*, Statist. Sci., 1, pp. 521–522.

G. WAHBA (1989), *On the dynamic estimation of relative weights for observation and forecast in numerical weather prediction*, Remote Sensing Retrieval Methods, A. Deepak, H. Fleming, and J. Theon, eds., A. Deepak Publishing, Hampton, VA, pp. 347–358.

G. WAHBA AND Y. WANG (1987), *When is the optimal regularization parameter insensitive to the choice of the loss function?*, Tech. Report 809, Department of Statistics, University of Wisconsin, Madison, WI.

G. WAHBA AND J. WENDELBERGER (1980), *Some new mathematical methods for variational objective analysis using splines and cross-validation*, Monthly Weather Rev., 108, pp. 1122–1145.

G. WAHBA AND S. WOLD (1975), *A completely automatic French curve*, Commun. Statist., 4, pp. 1–17.

W. WECKER AND C. ANSLEY (1983), *The signal extraction approach to non-linear regression and spline smoothing*, J. Amer. Statist. Assoc., 78, pp. 81–89.

E. WEGMAN AND I. WRIGHT (1983), *Splines in statistics*, J. Amer. Statist. Assoc., 78, pp. 351–366.

H. WEINERT, ED. (1982), *Reproducing kernel Hilbert spaces: Applications in Statistical Signal Processing*, Hutchinson Ross, Stroudsburg, PA.

H. WEINERT AND T. KAILATH (1974), *Stochastic interpretations and recursive algorithms for spline functions*, Ann. Statist., 2, pp. 787–794.

J. WENDELBERGER (1982), *Smoothing noisy data with multidimensional splines and generalized cross validation*, Ph.D. thesis, Department of Statistics, University of Wisconsin, Madison, WI.

E. T. WHITTAKER (1923), *On a new method of graduation*, Proc. Edinburgh Math. Soc., 41, pp. 63–75.

H. WOLTRING (1985), *On optimal smoothing and derivative estimation from noisy displacement data in biomechanics*, J. Human Movement Sci., 4, pp. 229–245.

H. WOLTRING (1986), *A* FORTRAN *package for generalized cross-validatory spline smoothing and differentiation*, Adv. in Engrg. Software, 8, pp. 104–113.

A. YAGLOM (1961), *Second order homogeneous random fields*, in Proc. Fourth Berkeley Symposium, J. Neyman, ed., University of California Press, Berkeley, CA, pp. 593–622.

T. YANAGIMOTO AND M. YANAGIMOTO (1987), *The use of marginal likelihood for a diagnostic test for the goodness of fit of the simple linear regression model*, Technometrics, 29, pp. 95–101.

B. YANDELL (1986), *Algorithms for nonlinear generalized cross-validation*, in Computer Science and Statistics: 18th Symposium on the Interface, T. Boardman, ed., American Statistical Association, Washington, DC, pp. 450–455.

Author Index